Designing an Innovative Pedagogy for Sustainable Development in Higher Education

Higher Education and Sustainability

J. Paulo Davim

Professor, Department of Mechanical Engineering, University of Aveiro, Portugal

This new series fosters information exchange and discussion on higher education for sustainability and related aspects, namely, academic staff and student initiatives, campus design for sustainability, curriculum development for sustainability, global green standards: ISO 14000, green computing, green engineering education, index of sustainability, recycling and energy efficiency, strategic sustainable development, sustainability policies, sustainability reports, etc. The series will also provide information on principles, strategies, models, techniques, methodologies, and applications of higher education for sustainability. It aims to communicate the latest developments and thinking as well as the latest research activity relating to higher education, namely, engineering education.

Higher Education and Sustainability
Opportunities and Challenges for Achieving Sustainable Development Goals
Edited by Ulisses Manuel de Miranda Azeiteiro and J. Paulo Davim

Designing an Innovative Pedagogy for Sustainable Development in Higher Education
Edited by Vasiliki Brinia and J. Paulo Davim

For more information about this series, please visit: www.crcpress.com/Higher-Education-and-Sustainability/book-series/CRCHIGEDUSUS

Designing an Innovative Pedagogy for Sustainable Development in Higher Education

Edited by
Vasiliki Brinia and J. Paulo Davim

CRC Press
Taylor & Francis Group
Boca Raton London New York

CRC Press is an imprint of the
Taylor & Francis Group, an **informa** business

by CRC Press
6000 Broken Sound Parkway NW, Suite 300, Boca Raton, FL 33487-2742

and by CRC Press
2 Park Square, Milton Park, Abingdon, Oxon, OX14 4RN

© 2020 Taylor & Francis Group, LLC
CRC Press is an imprint of Taylor & Francis Group, LLC

ISBN: 978-0-367-18225-0 (hbk)
ISBN: 978-0-429-06019-9 (ebk)

Typeset in Times
by codeMantra

Contents

Preface

"Just to be on the first step should make you happy and proud. To have reached this point is no small achievement: what you've done already is a wonderful thing…"

Constantine P. Cavafy, The first step, *1896–1904*

Sustainability or sustainable development is one of the most important issues concerning intergenerational equity and long-run fairness of the whole society, yet it has never taught in an experimental, artful, and design-based way in higher education for teaching our students how to design their future and the future of our planet in order to be sustainable.

One reason for this is that there is a gap between real-world problems and the current pedagogy provided by higher education systems. Also, in current educational systems, responsibility and creativity are not valued enough. As an example, in a sample of about 350, fourth-year students in Teacher Education Program (TEP) of Athens University of Economics and Business (AUEB), only a few students have had a learning experience of solving—in an experimental, creative, and collaborative way—a real-world problem. Furthermore, no one had used during their undergraduate studies, any arts-based or design-thinking learning practice.

However, when those students were taught during their studies in TEP-AUEB the experimental culture of thinking and learning by using collaborative and innovative approaches, the results had been proven to be really impressive. The artifacts that they created for solving real-world problems in an abstract, artistic, and hands-on way were really fantastic (see: https://goo.gl/JNQn1b).

This book is—as far as we know—the first one that intends to develop a pedagogy for sustainable development (PfSD) based on arts, design thinking, technology, and entrepreneurship. We believe that sustainability-oriented competences could be efficiently taught in higher education based on a holistic way of thinking by using an experimental, collaborative, and innovative way of learning. We hope that this book will inspire educators, researchers, and decision-makers to consider innovative methods and practices for teaching sustainability in higher education in order for our students to acquire future skills for critical thinking and personal, societal, and economic responsibility.

Editors

Dr. Vasiliki Brinia is the scientific director of the Teacher Education Program (TEP) of the Athens University of Economics and Business (AUEB) and a lecturer at the Department of Informatics, School of Information Sciences, and Technology of AUEB. She is the head of the scientific committee for the TEP of the Hellenic Open University. She holds two bachelor's degrees in Business Administration, a master's degree in Business Administration (MBA), a diploma in Pedagogical Studies, a diploma in Counseling, and a doctorate (Ph.D.) in Education and Teaching Sciences with specialization in the teaching methodology of secondary education with experiential teaching methods. She has carried out postdoctoral research in the teaching methodology of adult education with experiential teaching methods. She has been an expert-advisor at the Ministry of Education in the organizational and administrative restructuring of the Greek educational system. She has long-term administrative and teaching experience in secondary education, higher education, and adult education. She has pursued extensive research and writing activities in Greece and abroad. In particular, she has had over 80 research papers published in international scientific journals with the blind peer-reviewed system. Also, she has published more than 35 papers in conferences. She is an active member of many Editorial Advisory Boards as editor in chief, guest editor of journals, books editor, associate editor, and reviewer for many international journals and conferences. In addition, she has also published as the author (and co-author) five textbooks.

Dr. J. Paulo Davim received his Ph.D. degree in Mechanical Engineering in 1997, M.Sc. degree in Mechanical Engineering (materials and manufacturing processes) in 1991, degree in Mechanical Engineering (5 years) in 1986, from the University of Porto (FEUP); the aggregate title (Full Habilitation) from the University of Coimbra in 2005; and the D.Sc. from London Metropolitan University in 2013. He is a senior chartered engineer by the Portuguese Institution of Engineers with an MBA and specialist title in Engineering and Industrial Management. He is also the recipient of Eur Ing (European Engineer) award from FEANI-Brussels and is a Fellow of the Institution of Engineering and Technology (FIET) by IET-London. Currently, he is a professor at the Department of Mechanical Engineering of the University of Aveiro, Portugal. He has more than

30 years of teaching and research experience in Manufacturing, Materials, Mechanical, and Industrial Engineering, with special emphasis in Machining and Tribology. He has also interest in Management, Engineering Education, and Higher Education for Sustainability. He has guided large numbers of postdoctoral, Ph.D., and master's students. Also, he has coordinated and participated in several financed research projects. He has received several scientific awards. He has worked as an evaluator of projects for ERC (European Research Council) and other international research agencies as well as an examiner of Ph.D. thesis for many universities in different countries. He is the editor in chief of several international journals, guest editor of journals, books editor, book series editor, and scientific advisor for many international journals and conferences. Presently, he is an editorial board member of 30 international journals and acts as a reviewer for more than 100 prestigious Web of Science journals. In addition, he has also published as the editor (and co-editor) more than 120 books and as the author (and co-author) more than 10 books, 80 book chapters, and 400 articles in journals and conferences (more than 250 articles in journals indexed in Web of Science core collection/h-index 52+/9,000+ citations, SCOPUS/h-index 57+/11,000+ citations, Google Scholar/h-index 74+/17,000+).

Contributors Bio

Athanassios Androutsos is currently a Faculty Member at the Department of Informatics at Athens University of Economics and Business (AUEB). He teaches Information and Education Technologies at the undergraduate and postgraduate level. He also teaches software engineering and programming at the Lifelong Center of AUEB. His main research involves network economics, design thinking, computer programming, and education technology. He has participated in several national and European research projects. He is a member of the Programme Committee of Teacher Education Programme at AUEB (https://www.dept.aueb.gr/en/tep). His work has been published in high impact factor international journals such as *IEEE Journal on Selected Areas in Communications*, conferences' proceedings, and book chapters.

Dr. Vasiliki Brinia is the Scientific Director of the Teacher Education Program (TEP) of the Athens University of Economics and Business (AUEB) and a Lecturer at the Department of Informatics at the School of Information Sciences and Technology of AUEB. She is the Head of the Scientific Committee for the TEP of the Hellenic Open University. She holds two bachelor's degrees in Business Administration, a master's degree in Business Administration (MBA), a diploma in Pedagogical Studies, a diploma in Counseling, and a doctorate (Ph.D.) in Education and Teaching Sciences with specialization in the teaching methodology of secondary education with experiential teaching methods. She has carried out postdoctoral research in the teaching methodology of adult education with experiential teaching methods. She has been an expert-advisor at the Ministry of Education in the organizational and administrative restructuring of the Greek educational system. She has long-term administrative and teaching experience in secondary education, higher education, and adult education. She has pursued extensive research and writing activities in Greece and abroad. In particular, she has made 83 publications, which are research papers that have been published in international scientific journals with the blind peer-reviewed system. Also, she has published more than 35 papers in conferences. She is an active member of many Editorial Advisory Boards as editor in chief, guest editor of journals, books editor, associate editor, and reviewer for many international journals and conferences. In addition, she has also published as author (and co-author) for five textbooks.

Dr. Nikos Chatzistamoulou is an empirical economist by training. His research field is Applied Microeconometrics, and his research interests entail the fields of Efficiency and Productivity Analysis, Environmental & Energy Economics, and Economics of Sustainable Development. He holds a four-year B.Sc. in Economics from the University of Patras, Greece; M.Sc. in Economics from the Athens University of Economics and Business; and Ph.D. in Economics from the University of Patras. He has held research positions at the Athens University of Economics and Business, Greece; the University of Surrey, the United Kingdom; and the University of Patras, Greece. He has teaching experience both at the undergraduate and postgraduate level. He has co-authored two books in Applied Operations Research using R while his research has been published to international peer-reviewed journals (*Energy Economics, Water Journal, International Journal of Production Economics, Technological Forecasting and Social change,* and Environmental Economics and Policy Studies). His research has been funded by national and European projects as well.

Thalia Dragonas is a professor of Social Psychology, dean of the School of Educational Sciences at the National and Kapodistrian University of Athens, Greece. She has written extensively on psychosocial identities, intercultural education and ethnocentrism in the educational system. Since 1997 she is co-directing a large-scale intervention on the education of the Muslim Minority in Western Thrace. She co-edits a book series on Social and Historical Studies in Greece and Turkey (I.B. TAURIS). She was member of Greek Parliament (2007–2009) with the Socialist Party (PASOK) and special secretary of Intercultural Education at the Ministry of Education (2010).

Vasiliki Karampa is a Ph.D. student in the research field of Technology-Enhanced Learning (TELE). She has graduated from the Department of Informatics at Athens University of Economics and Business (AUEB), and she holds an M.Sc. in eLearning from the Department of Digital Systems at the University of Piraeus.

Her research is oriented to the Smart Learning Environments and Smart Pedagogy with the vision of producing effective educational solutions in order to help students augment their skills and become even "smarter", as to face and solve the humanity's problems. More specifically, her work encompasses studies regarding the education for sustainable development, especially in higher education. Furthermore, she investigates aspects of learning through STEAM education, Internet of Things (IoT), and robotics as well as programming into blended and e-learning settings.

She has participated in national and international conferences, and she has also been running STEAM/IoT workshops for students at the Department of Digital Systems at the University of Piraeus.

 **Professor Phoebe Koundouri** holds a Ph.D. and M.Phil. in Economics from the University of Cambridge (UK). She is a Professor (Chair) of Sustainable Development (Economics and Econometrics) at the School of Economics at Athens University of Economics and Business (Greece), and she is the elected President of the European Association of Environmental and Natural Resource Economists (https://www.eaere.org/) with more than 1200 scientific institutions as members from more than 60 different countries. Professor Koundouri is listed in the 1% of most-cited women economists in the world, with 15 published books and more than 250 published scientific papers.

Professor Koundouri is also the Founder and Scientific Director of the Research Laboratory – ReSEES: Research on Socio-Economic and Environmental Sustainability www.dept.aueb.gr/en/ReSEES at the Athens University of Economics and Business and an Affiliated Professor at the ATHENA Research and Innovation Center (Greece), where she directs EIT Climate KIC Hub Greece of the European Institute of Innovation and Technology (www.climate-kic.org). She is also the Co-Chair of the United Nations Sustainable Development Network – Greece (www. unsdsn.org), Chair of the Scientific Advisory Board of ICRE8 International Research Center, and Chair of the scientific advisory board of the European Forest Institute (www.efi.int).

In the past, Professor Koundouri has held academic positions at the University of Cambridge, the University College London, the University of Reading, and the London School of Economics. She acts as an advisor to the European Commission, World Bank, EIB, EBRD, OECD, UN, NATO, WHO, numerous national and international foundations and organizations, as well as national governments in all five continents. Notably, she is currently member of the drafting Priministerial Committee for 10-year development plan for Greece, as well as in the Climate Change Committee of the Greek Ministry of Environment and Energy. Since 1997, she has coordinated more than 65 interdisciplinary research projects and has attracted significant competitive research funding.

Professor Koundouri and her large interdisciplinary team have produced research and policy results that have contributed to accelerating the research commercialization for the sustainability transition in Europe, as well as shaping

European policies. Over the last two decades, Professor Koundouri has given keynote and public lecturers all over the world and received various prizes for academic excellence.

 **Eva Österlind, PhD** in Ed. Sc., Professor in Drama Education at Stockholm University, has studied drama in the Nordic curricula, and conducted comparative studies of Upper Secondary students' experiences of Drama. She has also analyzed the transformative potential of drama, using Bourdieu's concept of habitus, and discussed how evaluation of theater for social change could be improved. Her main field of teaching is teacher education and CPD courses for teachers. She designed the first drama courses at advanced level in Sweden and now leads Master Education in Drama and Applied Theatre. She also tutors doctoral students. Her research interests concern the potential of Drama in Education, Forum Theatre as a tool for active citizenship, and, last but not least, Drama in Higher Education for Sustainability.

 **Foteini Paraskeva** is an Associate Professor of Learning Psychology with Technology in the Department of Digital Systems (DDS) at the University of Piraeus, Greece. She received her Ph.D. in Educational Psychology in the domain of learning with technology applications. Her research interests focus on study learning and instruction with I.C.T. and e-learning-related settings to optimize leaning and performance, with particular emphasis on social, emotional, and cognitive aspects of human learning. Her publications focus on (e)-learning, smart pedagogy, self-regulated learning, self-efficacy, and motivational beliefs, as well as scenario-based learning in academic and business settings. She has published a significant work in conferences and journals, and she has been an active participant in many international scholarly groups and SIGs.

Yiannis Roussakis is an Assistant Professor of Pedagogy in the Department of Special Education at the University of Thessaly, where he teaches Comparative Education and Introduction to Pedagogy, and supervises the Teaching Practicum of student teachers. He holds a Ph.D. in Comparative Education and Education Policy (University of Athens), M.Ed. in Pedagogy, B.Sc. in Physics, and B.Ed. in Primary Education. He is the co-author of one book ("European Union: Policies for Education", Greek Pedagogical Institute, 2008) and co-editor of two collective volumes. He is also the author of many collective-volume chapters and scientific journal papers in Greek and in English on European education policies in comparative perspective, teacher education policy and practice, educational assessment and quality improvement, education for sustainable development, and Information and communication technologies (ICTs) in education.

Helen Salavou is currently an Associate Professor of Business Administration at the Athens University of Economics and Business. Her main research interests involve entrepreneurship (traditional and social), innovation, and strategy of small firms. She is the Founding Member of the research unit of social entrepreneurship (http://use.aueb.gr/) and the Scientific Director of the annual education program in social entrepreneurship (https://www.dose.aueb.gr/) at the university. She has published her work in international journals (~2,150 citations) and participated in several national and international research projects. Parallel to her academic career she has contributed as management consultant and mentor in numerous entrepreneurial initiatives.

Introduction

Sustainability has always been at the center of discussions worldwide, since the desired economic growth needs to be prudently combined with the finite natural resources of the planet that nurture humanity and safeguard its continuity. Even more in recent years, when the global community seems to have failed to effectively tackle the environmental challenges that arise, there has been an urgent call to reinvent sustainable ways of living and utilizing existing resources. The educational system is thought to hold great part of the responsibility for this phenomenon. For this reason, education is now seen as a key factor that can and should have a leading role, in order to help develop the appropriate skills and instill the needed values in young people, who are trying to "build" a sustainable present and a prosper future for the generations to come.

After the UN launched the 17 Sustainable Development Goals (SDGs), which constitute a guide to foster sustainability practices worldwide, most universities inaugurated several practices accordingly. However, a more systematic and holistic approach needs to be explored and implemented regarding education for sustainable development (ESD). Therefore, the aim of the present book is to discuss and propose innovative teaching and educational approaches that—if systematically implemented—can effectively address the issue of sustainability. An interdisciplinary approach to education and teaching is seen as the only way to develop holistic perspectives and methods that will serve the needs of ESD. The innovation of the present book lies exactly on the fact that sustainability, because of its complex and manifold character, can only be tackled through holistic, interdisciplinary methods that will be universally adopted in higher education. In this sense, the interdisciplinary approaches of the contributions made for the present book are briefly presented below.

In *Chapter 1*, the authors start the discussion by analyzing the term "sustainability economics." This term is crucial for the overall discussion of this book, and for this reason, the different definitions that have been developed through the years are cited and critically regarded, including not only the aspect of economic growth but also the aspect of prosperity for the future generations. A historic analysis of the term will shed light on the different aspects it embraces and will clarify the conditions for the discussion to follow.

In *Chapter 2*, the authors discuss the need to develop appropriate educational frameworks on project-based learning, which will be based on the constructivist method and "smart" learning approaches, in order for future generations to be able to cope with real-world problems. This framework is suggested to be linked with the interdisciplinary nature of STEM, so as to serve as the foundation for the development of academic programs and curricula which will enhance the graduates' skills on research and problem-solving for sustainable development.

Chapter 3 touches on the topics of how cultural heritage, drama, and art can be linked to and enhance sustainability through education. The first article of the chapter discusses the innovative topic of how teachers can include cultural heritage in the learning process in order to develop students' skills on sustainability through project-based learning and social entrepreneurship.

The second article of *Chapter 4* discusses how drama can be utilized as an art-based method in teaching, since it is based on the use of all students' senses, their own action, and reflection on their experience, and it can thus help them understand and assimilate knowledge on sustainability in an effective experiential way.

Chapter 5 is about the so-called knowledge-based economy and its relation to sustainable development. This is a topic of great importance for the present discussion, since new forms of knowledge need to be developed, in order to generate higher levels of economic growth without at the same time compromising the quality of life of the future generations. An economic model of knowledge creation and long-term sharing is therefore suggested in this chapter.

In *Chapter 6*, the author discusses about the skills of entrepreneurship and social entrepreneurship and how these can be developed in higher education as tools for sustainable development. The first article of the chapter constitutes an empirical study that explores differences between two groups of young people: one group having entrepreneurship education and the other one not having such an education.

The second article of *Chapter 7* describes an innovative course on social entrepreneurship taught in early childhood teacher education. This course aims at nurturing a mindset for entrepreneurship that will enable future teachers not only to instill entrepreneurship values to minors but also to identify themselves, to identify creative concepts, and to proceed to relevant action.

The last chapter of the book, *Chapter 8*, discusses the so-called education for sustainability (EfS) or else known as education for sustainable development (ESD)—as also discussed above—terms that came to the forefront of the field of education recently because of the failure of modern economies to effectively face the challenges of sustainable development. This failure is often seen by many as a failure of the educational system in general and, of higher education more specifically, to instill in young people the needed skills and principles, in order for them to become responsible citizens that will generate ideas of value for the future of the planet and of humanity. Design thinking, digital transformation, art-based teaching methods, and entrepreneurship for sustainable development are the aspects that need to be systematically included in educational practices for sustainability and that are discussed in this book.

As it becomes obvious through the brief presentation of the chapters of the book, the recent trend, but also the new reality with a universal truth and perspective, is the interdisciplinary approach in ESD. This approach takes lessons and experience from many fields to create a holistic and systematic method that needs to be gradually included in the curricula of current and future programs worldwide.

Higher education is thought to be one of the most significant factors that can and should contribute to shaping the way sustainability is seen nowadays by scientific researchers, practitioners, and the leading political community.

Dr Vasiliki Brinia
Teacher Education Programme, Scientific coordinator,
Department of Informatics, School of Information Sciences and Technology,
Athens University of Economics and Business, Greece

Dr Paulo Davim
Professor at the Department of Mechanical Engineering of the
University of Aveiro, Portugal

1 The Economics of Sustainable Development

Nikos Chatzistamoulou
Athens University of Economics and Business,
School of Economics and ReSEES Research Laboratory,
University of Patras

Phoebe Koundouri
Athens University of Economics and Business,
School of Economics and ReSEES Research Laboratory,

ATHENA Research and Innovation Center,
EIT Climate KIC Hub Greece,

UN Sustainable Development Solutions Network Greece

CONTENTS

1.1 INTRODUCTION

1.1.1 THE THEORY OF ECONOMIC GROWTH THROUGH BASIC MODELS

Within the Theory of Economic Growth, different strands have been surfaced throughout the years aspiring the explained variation in income levels across countries, or in other words, how nations grow and prosper. Solow-Swan model (1956) was a pioneering approach using aggregate production functions to study the accumulation of capital in order to improve nations' income. Although influential, Solow's model assumes that technical progress is exogenous. In other words, the model assumes that technological progress captured by the (unexplained by the model) Solow residual just happens. The latter is the main shortcoming of the model as in the absence of technological progress, sustained growth is questionable. Moreover, the rate of capital accumulation (i.e., change) is determined by the savings rate, the depreciation, and the rate of population growth, respectively, which are also assumed to be exogenous. In other words, in this exogenous growth model, the interest is placed on capital accumulation as this was believed to be the means for welfare and prosperity at that time, thus neglecting the role of technology in sustained growth. Nevertheless, Mankiw et al.'s model (1992) provides an estimate of the total factor productivity rate, that is, the Solow residual.

Another strand is the neoclassical growth model that explicitly determines consumer's side by taking into consideration (or in economic terms, endogenizing) savings. The Ramsey (1928) or Cass (1965)-Koopmans model (1965) introduces household optimization assuming an infinitely living representative household. Household preferences are specified; therefore, savings can now be linked to them, along with technology and prices in the economy. The most important contribution attached to this line of models is that it paves the way for a more systematic analysis of capital accumulation, investment in human capital, and endogenous technological progress. Although it does not shed light on the causes of cross-country income differences and economic growth, it clarifies the nature of economic decisions. However, the main assumption of the model is its major shortcomings as well.

In the overlapping generation models (OLGs), such as Diamond's model (1965), households do not live eternally, but also allow for new households in the economy over time; however, their usefulness is not limited to that. Totally different implications are derived compared to the aforementioned neoclassical model, while the dynamics of capital accumulation and consumption are closer to Solow's rather than the neoclassical model.

The common feature of the models described so far is the focus on exogenous technological progress as the source of capital accumulation and economic growth. However, the latter was challenged by the Endogenous Growth Theory stating that technical progress occurs within the system through Research and Development (R&D) activities and is therefore endogenous. The most important representatives of this class of models are David Romer and Robert Lucas.

On the one hand, Romer (1994) assumes that generated knowledge by a firm's research spills over to other ones, creating new knowledge for them as well. In other words, technology has spillover effects across the entire economy and constitutes the ultimate determinant of long-run growth. On the other hand, Lucas (1988) puts human capital under the spotlight. Investment in education contributes to the production of human capital boosting growth. He argues that through education, the individual worker undergoing training becomes more productive (internal effect) and that spillovers increase both the productivity of capital and that of other workers in the economy (external). It is an investment in human rather than physical capital that has spillover effects, thus increasing the level of technology.

Overall, the focus is on technological progress, without explicit explanation of the details of the investment *per se*. However, income differences across countries could be attributed to differences in technology levels. Thus, understanding the sources of such differences is a necessary condition to achieve economic growth.

Grossman and Helpman (1994) brought to the discussion the role of the innovation process to explain the growth by arguing that research leads to a greater variety of final goods, and income improves because households gain more utility through product proliferation. However, a country's technological progress is solely determined by its own investment in R&D, which is questionable. Technological advancements diffuse across countries, and each country has the potential to absorb knowledge generated through the World Technology Frontier, thus making diffusion equally important to the creation of new technologies. Another limitation of these models is that they do not capture the notion of the creative destruction process; that is, although innovation creates new technologies, it also "destroys" others by making them obsolete. Last but not least, the Schumpeterian models of economic growth capture process have their own limitations that go beyond the scope of this chapter.

1.1.2 THE ROLE OF TECHNOLOGY IN PRODUCTION

A well-known fact in economic theory is that the production functions, that is, production frontiers of economies of two different countries, are not directly comparable due to differences in the level of technology they have access to. Empirically speaking, this is one of the main issues that cause inconvenience to researchers in cross-country comparisons using most of the times the gross domestic product. A production function combines inputs such as labor, capital, and energy in order to produce outputs such as gross domestic product by assuming a certain level of technology. The latter is not directly observable, and it is considered as a black box.

Recent methodological advancements acknowledge that technology is a source of productivity and prosperity differences revolutionized the way cross-nation comparisons and benchmarking are done. The pioneering work of Haymi (1969) and Hayami and Ruttan (1970) introduced the concept of the metaproduction function that envelops all the individual frontiers. Moreover, the influential contribution of O'Donnell et al. (2008) further developed the concept and notions that could be used for benchmarking purposes and performance evaluation of the decision-making units under examination.

The latter paves the way for calculating the technology gap, which is the distance between the individual frontier and the metafrontier. The metafrontier that is used as an empirical tool to account for all the possible heterogeneities among the units is under consideration, has become a growing wave, and has triggered many studies in the field of economics in terms of efficiency and productivity.

It has become apparent that growth and prosperity can be sustained through technology diffusion. Recent contributions of Tsekouras et al. (2016) and Chatzistamoulou et al. (2019) have used the metafrontier to capture knowledge flows; that is, spillover effects improve the performance of the units bringing to the forefront, and productivity differences are attributed to heterogeneities of technology.

1.2 CONCEPTUAL FRAMEWORK

1.2.1 TRACING THE EVENTS: A HISTORICAL BACKGROUND

In 1982, the term "sustainability" enters the scene in the World Charter for Nature (United Nations—UN, 1982), but the concept was officially introduced in the Brundtland Report (WCED, 1987), which was the launchpad of the emergence of the sustainability literature (Pezzey & Toman, 2002). However, there was an earlier contribution by Barbier (1987). It was elaborated in the Agenda 21 during the Earth Summit in 1992 (UN, 1992).

The social dimension of sustainability was first introduced at the World Summit on Social Development in Copenhagen in 1995 (UN, 1995) and later endorsed by the World Summit in Johannesburg in 2002 (UN, 2002). Prior to that, during the Millennium Summit (UN, 2000), the Millennium Development Goals (MDGs) surfaced from the Millennium Declaration as a set of eight goals for the period 2000–2015, while the Conference in Rio De Janeiro in 2012 embraced the outcome of 2002 Summit (UN, 2012b), namely, the social pillar of sustainability, receiving significant attention.

The Sustainable Development Goals (SDGs) was initially proposed in 2014 by the Open Working Group of the UN General Assembly (UN, 2014) and was only adopted in 2015 (UN, 2015), as a network of goals (Le Blanc, 2015) that is important to understand (Singh et al., 2018). SDGs will be achieved by 2030 to inherit the MDGs. The initiative comprises 17 goals, 169 targets, and a few hundreds of indicators.

1.2.2 LITERATURE REVIEW

Although sustainability has become a modern buzzword, any attempt to provide a formal definition is not an easy task, mostly due to the fact that it encapsulates many concepts, meanings, and perspectives. It has become already apparent that early studies have placed an interest on how economic growth occurs; the capital accumulation; and then the role of technology and human capital deepening and innovation. However, for achieving sustainable development, all of the aforementioned parameters are necessary, but not sufficient.

In the beginning, "sustainability" has mainly had an environmental-ecological flavor. The term "sustainability" is considered as a normative notion that is related to the human-nature relationship and the current generation-next generation relationship. Although a coherent consensus about the precise content of "sustainability economics" is not readily available, a growing body of literature suggests that it aims at efficient management considering uncertainty entailing a cognitive interest as well. This is what makes "sustainability economics" a relevant science (Baumgärtner & Quaas, 2010—BQ), and it has also been linked to a final product (Victor, 1991).

In developing the term, Van den Bergh (2010), through a critique on the work of BQ, suggests that environmental externalities and sustainability should be conceptually associated, highlights other facets of sustainability (e.g., weak, strong, spatial), and argues that offering policy insights is at the core of the field. Baumgärtner & Quaas, 2010 and Van den Bergh (2010) have triggered an active discussion bringing to the forefront the origins of sustainability concept derived from the Summits, whereas Bartelmus (2010) argues about the usefulness of the field. Other important contributions to the concept of sustainable development/sustainability are those by Daly (1990), Pezzey (1992), Toman (1994), and Beckerman (1994) just to mention a few. Although it has been argued that the two terms differ, in fact those embrace the identical facets and lead to similar policy suggestions. It should be noted that Pezzey and Toman (2002) enlighten our understanding through a systematic literature review about sustainability economics.

Therefore, the terms are used interchangeably by highlighting the long-run orientation. Nowadays, due to the intrinsic complexity and multidisciplinary content of sustainability, no clear definition exists to serve as a guideline for policy making (Holden et al., 2014). Research has increased awareness around the concept, indicating that it is a rather multidimensional and by all means not a straightforward notion.

Ecosystem services are also related to sustainability and have also gained a merit in the body of literature. In particular, efficient resource management, and more precisely, the water management of all kinds, has been extensively studied as its scarcity hinders sustainable development across the globe. The Water Framework Directive (WFD) sets the guidelines for efficient management, and in this line, Koundouri et al. (2016) introduce a methodology that evaluates the total economic value of water services aligned with the WFD. Navarro-Ortega et al. (2015) and Akinsete et al. (2019) under the GLOBAQUA project explore the linkages between the factors affecting water quality and the factors affecting human well-being by focusing on several river basins, respectively. Sustainable river management (Dávila et al., 2017; Pistocchi et al., 2017); the value of biodiversity (Birol et al., 2009); the marine and coastal ecosystems mitigation measures against climate change (Remoundou et al., 2009, 2015), the oceans (Koundouri, 2017; Koundouri & Giannouli, 2015), and the seas (Remoundou et al., 2014; Stuiver et al., 2016; Van den Burg et al., 2016; Zagonari et al., 2018) have also been subject to research.

1.2.3 MEASUREMENT EFFORTS

The interest in monitoring sustainability over time generates the urge for proper measurement. A growing body of literature underlines the need for appropriate indicators, which are comparable across nations and universally accepted (e.g., Dahl, 2012; Hák et al., 2016). The SDGs could put the world in a *sustainable trajectory* (Sachs, 2012) as the former introduced as an opportunity to rationalize the expectations of their predecessors, that is, the MDGs (Joshi et al., 2015). Soon after the introduction of the former which made possible the measurement of sustainability, many studies focusing on specific goals have surfaced.

For instance, Le Blanc (2015) employs a network analysis to study the interlinkages between the SDGs focusing on the SDGs 10 and 12. The network analysis supports the links coming from targets that are listed under SDGs other than the selected ones, underlining the fact that those indeed constitute a network. Singh et al. (2018) also study the interconnections of SDGs, argue that they could drive the sophisticated policy making, and focus on SDG 14 (oceans) to highlight its importance in achieving sustainable development. Although the interest is placed on specific SDGs, studies analyzing the SDG index (Sachs et al., 2016, 2017, 2018, 2019) have yet to be surfaced.

Before the SDGs and the development of the SDG index (Sachs et al., 2015, 2016, 2017, 2018, 2019) and since sustainability was perceived to encapsulate environmental as well as human well-being concerns, corresponding indices, although partial, have been employed in empirical research. Just to mention a few of the most recent ones, Moran et al. (2008) use the UN Human Development Index (HDI) and the ecological footprint to measure SD within the ecological limits, by setting thresholds, to find that low-income countries managed to reduce the latter and increase the former, while the opposite holds true for the high-income countries.

Among the indices that have been employed, Dahl (2012) finds the Environmental Vulnerability Index, the Environmental Sustainability Index, and the Environmental Performance Index although it has been argued that it is not straightforward to assess sustainability adequately. Moreover, Holden et al. (2014) employ the HDI to measure equity representing quality as well as the Gini index measuring equality representing quantity by assigning thresholds as well.

This somehow foreshadows the SDG index as by Sachs et al. (2015, 2016, 2017, 2018, 2019) offering a unified framework for measuring sustainability achievement through the common set of goals, targets, and indicators facilitating cross-nation SD comparisons, by passing the need to set thresholds to determine the improvement. The latter, however, is one of the main points of criticism as different nations have different initial conditions affecting their progress.

All in all, the SDGs have revolutionized the way of integrating the same policy targets into the national-level discussion. Such an undertaking is most certainly multifaceted, and the use of the SDG index facilitates such comparisons. However, national-level political decisions prove to be hard to quantify and will remain unobserved. The main challenge is to promote the adoption and find a monitoring mechanism in an attempt to guide universal implementation.

1.2.4   AN INTRODUCTION TO THE CIRCULAR ECONOMY

We now shift the attention to the concept and meaning of the circular economy as it is a trending topic of the public dialogue. From the UN to the European Commission, both sides of the ocean have dedicated a significant amount of resources to communicate the importance of building a circular economy rather than wasting irrationally scarce resources to deliver a sustainable future to the generations to come. Thus, the aim is to put things in a perspective rather than offering an in-depth analytical framework.

The (First) Industrial Revolution that occurred in the 18th century set the foundation on which the modern (manufacturing mostly) processes are established. Ever since, the list with the events that occurred—that is, the technological progress—and changed the way production takes place is quite long. However, at that time, the philosophy of production was of a *linear form*, in the sense that raw materials, that is, inputs, were transformed into outputs with some sort of technologies, whereas nonrecyclable waste was occurring as a natural consequence. Time passed by and societies come to realize that this "take-make-waste" model is not a sustainable forward-looking strategy as it was exponentially draining the Earth's scarce resources to cover the needs of the modern societies.

Another notion related to that of the circular economy is the *recycle-reuse economy*. This is mainly based on the act of reusing or taking materials that have already been processed and adding value to them by generating new ones without involving additional raw materials. From the viewpoint of the economist, this is translated into generating value through products that have been produced and used previously so as to enter the system again, improving the prosperity and welfare of societies with restricted access to production means. However, it is also linked to higher quality of life since it comes with the creation of jobs and business opportunities. Nevertheless, the common feature of a linear economy and a reuse economy is that both are associated with nonrecyclable waste.

Undeniably, the term *circular economy* has been the buzzword that has dominated the public interest reaching a diversified audience, from academics and scholars to practitioners, stakeholders, and policy makers. This is mainly the reason it has received multiple meanings and content.

Although it is a very complex and tangled notion as it is affected by many streams of thought and disciplines (Ellen MacArthur Foundation, 2012, 2013), a rather general definition could be that *a circular economy is an economy that aims to extract fewer scarce resources as time goes by in order to minimize the waste produced by maintaining the value of the items for longer and putting them into the production cycle generating a feedback process to avoid using raw materials* (Eurostat, 2019).

Efficient use of resources and keeping waste to the bare minimum are central to the concept of the circular economy as a sophisticated product design and material use lead to a higher-quality product that is associated with less efforts to manage waste, preserve and respect the relative scarcity of resources, and generate value through the associated business opportunities that arise during the process. In short, *circular economy is a redesigning, regenerative, and restorative system towards the use of renewable energy* (Word Economic Forum, 2019).

However, the aforementioned is only a broad term regarding the circular economy. Recently, Kirchherr et al. (2017) conducted a systematic analysis regarding the definitions of the circular economy used in the current discourse. The findings indicate that there are more than 100 definitions, which may be a source of misunderstanding among the stakeholders and agents. They argue that some of the definitions adopted mix circular economy with recycling and that the links with sustainable development appear to be weak as the associations with the prosperity of the future generations are hardly part of the picture. The latter highlights the need for a clear understanding of the notion to boost its importance for the quality of life. Figure 1.1 illustrates the flow of process and the main differences between the concepts described above.

The implementation of the concept of the circular economy has led to the creation of the circular economy business models (CBMs). In particular, the CBM is a model that implements the principles underpinned by the circular economy philosophy. The latter may take the form of industrial symbiosis, the sharing economy, which is becoming especially popular in Europe. The circular-sustainable design is gaining ground on a worldwide scale, the reverse logistics and remanufacturing.

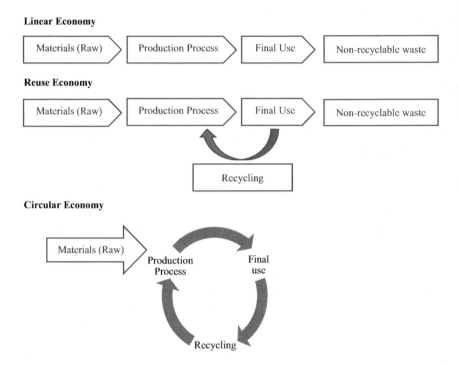

FIGURE 1.1   The linear, reuse, and circular economy. (Own construction.)

Irrespective of the CBM adopted and implemented, the benefits take a similar form. Above all is the effort to reduce the environmental footprint and reduce overloading the ecosystems from the effects of the production process while preserving the scarce resources such as energy and raw materials, boosting economic growth and competitiveness at a global scale, triggering innovation activity, and generating jobs that have proven to be among the main benefits.

It goes without saying that the research around circular economy proliferates in an exponential manner during the last decade. For instance, Ritzén and Sandström (2017) argue that the concept of the circular economy is not far from its complete implementation in practice as there are financial, structural, operational, attitudinal, and technological obstacles making the transition hard to be achieved. In a similar vein, Korhonen et al. (2018) report six potential challenges of the circular economy with respect to the sustainability of the environment: *the thermodynamic limits, the limits posed by physical scale of the economy, the limits posed by path dependency and lock-in, the limits of governance and management, the limits of social and cultural definitions, and the definition of physical flows.*

In addition to those challenges, it has been argued that the concept of circular economy does not encapsulate the concepts of social and environmental justice to be part of the future agenda, and it has also been highlighted the fact that the CBMs are not in a position to offer optimal solutions. However, the systematic review of the literature conducted by Kalmykova et al. (2018) identified that market-ready solutions to be applied already exist for the implementation levels of the adopted strategies.

The circular economy concept has evolved throughout the years to include more aspects that promote a more resilient and resource efficiency society. Currently, there are ten principles or R's that describe the concept of circular economy system and are therefore linked to each other in a feedback loop: refuse (R0), reduce (R1), resell/reuse (R2), repair (R3), refurbish (R4), remanufacture (R5), repurpose (R6), recycle (R7), recover (R8), and re-mine (R9) (Reike et al., 2018). Reike et al. (2018) offer a graphical representation of the system as well.

Despite a rapidly growing wave of research related to the concept of circular economy and sustainability, the exact content of the two notions is not easy to be defined. Prieto-Sandoval et al. (2018) provide a systematic literature review using content analysis to clarify the concept of circular economy. An interesting finding is that they mention that the multiplicity of the definitions stems from the interdisciplinary nature of the term as many scientific fields such as ecology, economics, engineering, design, and business have incorporated the notion among their analytic tools. They provide a timeline from the first industrial revolution to the commonly accepted viewpoint that circular economy is a bring on the wall to achieve sustainable development (Xue et al., 2010; Ma et al., 2014; Geissdoerfer et al., 2017; Kirchherr et al., 2017). Sustainability itself is not a straightforward term to define, though, as it has a few 100 definitions (Johnston et al., 2007), as already mentioned.

It should also be mentioned that circular economy and sustainability are linked through the inclusion of the former within the initiatives of SDGs. More precisely, the circular economy is found in the SDG 8, which promotes inclusive and sustainable economic growth, employment, and decent work (Decent Work and Economic growth; indicator 8.4 Sustainable Consumption and Production); in the SDG 9, which builds resilient infrastructure, promotes sustainable industrialization, and fosters innovation (Industry, Innovation, and Infrastructure; indicator 9.4 Green Industry); and in the SDG 12, which addresses resource and energy efficiency (Responsible Consumption and Production; indicators 12.2 Natural Resource Management and 12.5 Waste Management) (United Nations, 2019).

Geissdoerfer et al. (2017) offer a systematic, comprehensive, and illuminating presentation of, and the differences as well as similarities between, the two notions. Regarding the similarities, they identify that 12 of them are related to a wide range of interactions and interdependencies between the agents and the environment, such as the role of business innovation and technology, the interdisciplinary nature of the fields, and the necessity that many stakeholders need to cooperate among others. However, as far as the differences are concerned, they compare the two notions based on the origins of the term, the goals and motivation, potential benefits, and responsibilities, among others.

1.3 METHODOLOGY

1.3.1 INTRODUCING THE SUSTAINABLE DEVELOPMENT GOALS

The SDGs have been introduced by the UN in 2015 aiming to develop a framework of 17 interconnected goals—each proxied by additional targets, for monitoring the growth, economic prosperity, challenges, inequality, poverty, peace, climate change, responsible consumption and production—and to raise environmental awareness among others, so each goal and every target will be achieved on a global scale by 2030 not only for industrialized nations but also for developing and emerging nations.

Each goal has several targets that capture aspects of its content. Those targets are subject to change to achieve a better grasp of each SDG as time goes by. It is not worthless to mention that although the SDGs have been agreed by all countries, their implementation is not obligatory. The interested reader may find useful information about the initiatives of SDGs at the official site of UN SDGs.[1] As there is diversity in the aspects that need to be monitored, there are 17 goals of SDGs: no poverty (Goal 1); zero hunger (Goal 2); good health and well-being (Goal 3); quality education (Goal 4); gender equality (Goal 5); clean water and sanitation (Goal 6); affordable and clean energy (Goal 7); decent work and economic growth (Goal 8); industry, innovation, and infrastructure (Goal 9); reduced inequalities (Goal 10); sustainable cities and communities (Goal 11); responsible consumption and production (Goal 12); climate action (13); life below water (14); life on land (Goal 15); peace, justice, and strong institutions (Goal 16); and partnerships (Goal 17).

1.3.2   THE SDG INDEX

The SDG Index Report provides insight regarding the performance of nations with respect to the SDGs. It is produced by Bertelsmann Stiftung with the support of the Sustainable Development Solutions Network (SDSN) Secretariat and member institutions (UN, 2019) by bringing together information from official sources such as the World Bank, the Organization for Economic Co-operation and Development, national authorities, and research centers. It captures the average performance of the nations on all of the SDGs.

Aspiring to get the big picture as regards the SDG performance of nations, we devise a dataset by collecting, combing, and matching the most recent information on the SDG index from 2016 through 2019 (Sachs et al., 2016, 2017, 2018, 2019), including including 193 countries worldwide. Due to missing data, 45 countries have been excluded from the sample. Thus the final dataset is a balanced panel consisting of 148 countries over 4 years. Therefore, the panel consists 592 observations.

1.4   RESULTS AND DISCUSSION

Table 1.1 presents the basic descriptive statistics for the SDG index for the period of study. The index is increasing (on average) within the period, which is an encouraging fact as countries improve their performance in the targets and consequently at the SDGs. The minimum score is also being improved as time goes by and this is promising as well.

Figure 1.2 illustrates how the SDG index is distributed on a global scale for the period of study. The vertical line corresponds to the global average, for the period of study, indicating how well countries perform. About 56.08% of the sample perform

---

[1] Please follow www.un.org/sustainabledevelopment/.

TABLE 1.1

Basic Descriptive Statistics for the Sustainable Development Index, 2016–2019

Years	Mean SDG Index	St. Dev	Min SDG Index	Max SDG Index
2016	58.43	13.84	26.10	84.53
2017	64.94	10.97	36.70	85.60
2018	65.04	10.32	37.70	85.00
2019	66.26	10.22	39.08	85.22
Sample period	63.67	11.81	26.10	85.60

Source: Own construction.

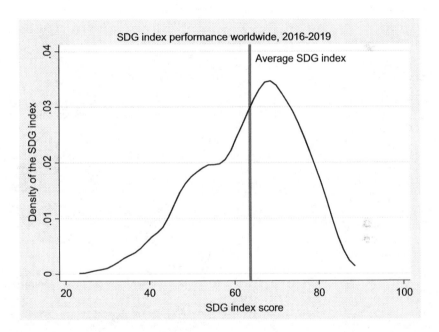

FIGURE 1.2   The distribution of SDG index globally, 2016–2019.

Source: Own construction

better than the average. Therefore, more than half of the sample exhibit an encouraging performance, indicating that most countries embrace and implement the SDGs to a satisfying extent.

On a final remark and as the scope of the chapter is to introduce the concepts, the interested reader may seek additional information through the UN SDG index and Dashboards official site, which provides amazing interactive tools and graphs about the countries' profiles.

1.5 CONCLUDING REMARKS

In its infancy, the concept of sustainable development was almost exclusively about the environmental-ecological footprint of societies related to the quality of the environment and the current generation that bequeaths to the next one, but progressively, governments, agencies, and authorities realized that there are most aspects included in the term. Nowadays, the term embraces a spherical appreciation of the human-environment relationship requiring an interdisciplinary approach in order to be studied.

Recent endeavors focus on the importance of the SDGs mostly by selecting a particular goal to study. The SDG index is gaining ground in the empirical analysis but research has not explored its potential yet. Combining it with other performance measures could offer useful insight into policy making. In light of the severe climate change, nations are becoming increasingly aware of the potential of the circular economy principles. The responsible authorities, however, should dedicate resources to increase awareness to the level of stakeholders and households to achieve the desired outcome.

REFERENCES

Akinsete, E., Apostolaki, S., Chatzistamoulou, N., Koundouri, P., & Tsani, S. (2019). The link between ecosystem services and human wellbeing in the implementation of the European water framework directive: Assessing four river basins in Europe. *Water*, *11*(3), 508.

Barbier, E. B. (1987). The concept of sustainable economic development. *Environmental Conservation*, *14*(2), 101–110.

Bartelmus, P. (2010). Use and usefulness of sustainability economics. *Ecological Economics*, *69*(11), 2053–2055.

Baumgärtner, S. & Quaas, M. (2010). What is sustainability economics? *Ecological Economics*, *69*(3), 445–450.

Beckerman, W. (1994). 'Sustainable development': is it a useful concept?. Environmental Values, *3*(3), 191–209.

Birol, E., Hanley, N., Koundouri, P., & Kountouris, Y. (2009). Optimal management of wetlands: Quantifying trade-offs between flood risks, recreation, and biodiversity conservation. *Water Resources Research*, *45*(11).

Cass, D. (1965). Optimum growth in an aggregate model of capital accumulation. *The Review of Economic Studies*, *32*(3), 233–240.

Chatzistamoulou, N., Kounetas, K., & Tsekouras, K. (2019). Energy efficiency, productive performance and heterogeneous competitiveness regimes. Does the dichotomy matter? *Energy Economics*, *81*, 687–697.

Dahl, A. L. (2012). Achievements and gaps in indicators for sustainability. *Ecological Indicators*, *17*, 14–19.

Daly, H. E. (1990). Toward some operational principles of sustainable development. *Ecological economics*, *2*(1), 1–6.

Dávila, O. G., Koundouri, P., Pantelidis, T., & Papandreou, A. (2017). Do agents' characteristics affect their valuation of 'common pool' resources? A full-preference ranking analysis for the value of sustainable river basin management. *Science of the Total Environment*, *575*, 1462–1469.

Diamond, P. A. (1965). National debt in a neoclassical growth model. *The American Economic Review*, *55*(5), 1126–1150.

Ellen MacArthur Foundation. (2012). *Towards the Circular Economy: Economic and Business Rationale for an Accelerated Transition.* Cowes: Ellen MacArthur Foundation.
Ellen MacArthur Foundation. (2013). *Towards the Circular Economy: Opportunities for the Consumer Goods Sector.* Cowes: Ellen MacArthur Foundation.
Eurostat, *Circular Economy,* accessible via https://ec.europa.eu/eurostat/web/circular-economy. Last accessed on 25 September 2019.
Geissdoerfer, M., Savaget, P., Bocken, N. M., & Hultink, E. J. (2017). The circular economy: A new sustainability paradigm? *Journal of Cleaner Production, 143,* 757–768.
Grossman, G. M. & Helpman, E. (1994). Endogenous innovation in the theory of growth. *Journal of Economic Perspectives, 8*(1), 23–44.
Hák, T., Janoušková, S., & Moldan, B. (2016). Sustainable development goals: A need for relevant indicators. *Ecological Indicators, 60,* 565–573.
Hayami, Y. (1969). Sources of agricultural productivity gap among selected countries. *American Journal of Agricultural Economics, 51*(3), 564–575.
Hayami, Y. & Ruttan, V. W. (1970). Agricultural productivity differences among countries. *The American Economic Review, 60*(5), 895–911.
Holden, E., Linnerud, K., & Banister, D. (2014). Sustainable development: Our common future revisited. *Global Environmental Change, 26,* 130–139.
Johnston, P., Everard, M., Santillo, D., & Robèrt, K. H. (2007). Reclaiming the definition of sustainability. *Environmental Science and Pollution Research International, 14*(1), 60–66.
Joshi, D. K., Hughes, B. B., & Sisk, T. D. (2015). Improving governance for the post-2015 sustainable development goals: Scenario forecasting the next 50 years. *World Development, 70,* 286–302.
Kalmykova, Y., Sadagopan, M., & Rosado, L. (2018). Circular economy: From review of theories and practices to development of implementation tools. *Resources, Conservation and Recycling, 135,* 190–201.
Kirchherr, J., Reike, D., & Hekkert, M. (2017). Conceptualizing the circular economy: An analysis of 114 definitions. *Resources, Conservation and Recycling, 127,* 221–232.
Koopmans, T. (1965). On the Concept of Optimal Growth, The Econometric Approach to Development Planning. Econometric Approach to Development Planning, 1st edn. North Holland, Amsterdam, 225–287.
Korhonen, J., Honkasalo, A., & Seppälä, J. (2018). Circular economy: The concept and its limitations. *Ecological Economics, 143,* 37–46.
Koundouri, P. (Ed.). (2017). *The Ocean of Tomorrow.* Cham: Springer International Publishing.
Koundouri, P. & Giannouli, A. (2015). Blue growth and economics. *Frontiers in Marine Science, 2,* 94.
Koundouri, P., Rault, P. K., Pergamalis, V., Skianis, V., & Souliotis, I. (2016). Development of an integrated methodology for the sustainable environmental and socio-economic management of river ecosystems. *Science of the Total Environment, 540,* 90–100.
Le Blanc, D. (2015). Towards integration at last? The sustainable development goals as a network of targets. *Sustainable Development, 23*(3), 176–187.
Lucas, R. (1988). On the mechanics of economic development. *Journal of Monetary Economics, 22,* 3–42.
Ma, S. H., Wen, Z. G., Chen, J. N., & Wen, Z. C. (2014). Mode of circular economy in China's iron and steel industry: A case study in Wu'an city. *Journal of Cleaner Production, 64,* 505–512.
Mankiw, N. G., Romer, D., & Weil, D. N. (1992). A contribution to the empirics of economic growth. *The Quarterly Journal of Economics, 107*(2), 407–437.
Moran, D. D., Wackernagel, M., Kitzes, J. A., Goldfinger, S. H., & Boutaud, A. (2008). Measuring sustainable development: Nation by nation. *Ecological Economics, 64*(3), 470–474.

Navarro-Ortega, A., Acuña, V., Bellin, A., Burek, P., Cassiani, G., Choukr-Allah, R., ... & Grathwohl, P. (2015). Managing the effects of multiple stressors on aquatic ecosystems under water scarcity. The GLOBAQUA project. *Science of the Total Environment, 503,* 3–9.

O'Donnell, C. J., Rao, D. P., & Battese, G. E. (2008). Metafrontier frameworks for the study of firm-level efficiencies and technology ratios. *Empirical Economics, 34*(2), 231–255.

Pezzey, J. (1992). *Sustainable Development Concept: An Economic Analysis.* The World Bank.

Pezzey, J. C. & Toman, M. (2002). The economics of sustainability: A review of journal articles (No. 1318–2016–103489).

Pistocchi, A., Udias, A., Grizzetti, B., Gelati, E., Koundouri, P., Ludwig, R., ... & Souliotis, I. (2017). An integrated assessment framework for the analysis of multiple pressures in aquatic ecosystems and the appraisal of management options. *Science of the Total Environment, 575,* 1477–1488.

Prieto-Sandoval, V., Jaca, C., & Ormazabal, M. (2018). Towards a consensus on the circular economy. *Journal of Cleaner Production, 179,* 605–615.

Ramsey, F. (1928). A mathematical theory of savings. *The Economic Journal, 38,* 543–559.

Reike, D., Vermeulen, W. J., & Witjes, S. (2018). The circular economy: New or refurbished as CE 3.0.? Exploring controversies in the conceptualization of the circular economy through a focus on history and resource value retention options. *Resources, Conservation and Recycling, 135,* 246–264.

Remoundou, K., Koundouri, P., Kontogianni, A., Nunes, P. A., & Skourtos, M. (2009). Valuation of natural marine ecosystems: An economic perspective. *Environmental Science and Policy, 12*(7), 1040–1051.

Remoundou, K., Adaman, F., Koundouri, P., & Nunes, P. A. (2014). Is the value of environmental goods sensitive to the public funding scheme? Evidence from a marine restoration programme in the Black Sea. *Empirical Economics, 47*(4), 1173–1192.

Remoundou, K., Diaz-Simal, P., Koundouri, P., & Rulleau, B. (2015). Valuing climate change mitigation: A choice experiment on a coastal and marine ecosystem. *Ecosystem Services, 11,* 87–94.

Romer, P. M. (1994). The origins of endogenous growth. *Journal of Economic perspectives, 8*(1), 3–22.

Ritzén, S. & Sandström, G. Ö. (2017). Barriers to the circular economy: Integration of perspectives and domains. *Procedia CIRP, 64,* 7–12.

Sachs, J. D. (2012). From millennium development goals to sustainable development goals. *The Lancet, 379*(9832), 2206–2211.

Sachs, J., Schmidt-Traub, G., Kroll, C., Durand-Delacre, D., & Teksoz, K. (2015). *An SDG Index and Dashboards: Global Report.* New York: Bertelsmann Stiftung and Sustainable Development Solutions Network (SDSN).

Sachs, J., Schmidt-Traub, G., Kroll, C., Durand-Delacre, D., & Teksoz, K. (2016). *An SDG Index and Dashboards: Global Report.* New York: Bertelsmann Stiftung and Sustainable Development Solutions Network (SDSN).

Sachs, J., Schmidt-Traub, G., Kroll, C., Durand-Delacre, D., & Teksoz, K. (2017). *SDG Index and Dashboards Report 2017.* New York: Bertelsmann Stiftung and Sustainable Development Solutions Network (SDSN).

Sachs, J., Schmidt-Traub, G., Kroll, C., Lafortune, G., Fuller, G. (2018). *SDG Index and Dashboards Report 2018.* New York: Bertelsmann Stiftung and Sustainable Development Solutions Network (SDSN).

Sachs, J., Schmidt-Traub, G., Kroll, C., Lafortune, G., Fuller, G. (2019). *Sustainable Development Report 2019.* New York: Bertelsmann Stiftung and Sustainable Development Solutions Network (SDSN).

Singh, G. G., Cisneros-Montemayor, A. M., Swartz, W., Cheung, W., Guy, J. A., Kenny, T. A., & Sumaila, R. (2018). A rapid assessment of co-benefits and trade-offs among sustainable development goals. *Marine Policy, 93,* 223–231.

Solow, R. M. (1956). A contribution to the theory of economic growth. *The Quarterly Journal of Economics, 70*(1), 65–94.

Stuiver, M., Soma, K., Koundouri, P., Van Den Burg, S., Gerritsen, A., Harkamp, T., ... & Hommes, S. (2016). The Governance of multi-use platforms at sea for energy production and aquaculture: Challenges for policy makers in European seas. *Sustainability, 8*(4), 333.

Swan, T. W. (1956). Economic growth and capital accumulation. *Economic Record, 32*(2), 334–361.

Toman, M. A. (1994). Economics and" sustainability": balancing trade-offs and imperatives. *Land Economics, 70*(4), 399–413.

Tsekouras, K., Chatzistamoulou, N., Kounetas, K., & Broadstock, D. C. (2016). Spillovers, path dependence and the productive performance of European transportation sectors in the presence of technology heterogeneity. *Technological Forecasting and Social Change, 102,* 261–274.

UN. (1982). A World Charter for Nature. United Nations, New York.

UN. (1992). Earth Summit Agenda 21. The United Nations programme of action from Rio. United Nations Department of Public Information, New York.

UN. (1995). Copenhagen Declaration on Social Development. World Summit for Social Development, held in March 1995 in Copenhagen, Denmark. www.un.org/.

UN. (2000). United Nations Millennium Declaration. United Nations General Assembly, 6–8 September 2000, United Nations, New York.

UN. (2002). Report of the World Summit on Sustainable Development. Johannesburg, South Africa, 26 August–4 Sept 2002. United Nations, New York.

UN. (2012a). The Future We Want. Resolution adopted by the general assembly on 27 July 2012, 66/288. United Nations.

UN. (2012b). Realizing the Future We Want for All. Report to the Secretary-General. UN System Task Team on the Post-2015 UN Development Agenda, New York.

UN. (2014). Millennium Development Goals Report 2014. United Nations, New York.

UN. (2015). Transforming Our World: The 2030 Agenda for Sustainable Development, 25–27 September 2015, United Nations, New York.

United Nations, *The Sustainable Development Goals,* accessible via https://sustainable-development.un.org/?menu=1300. Last accessed on 25 September 2019.

Van den Bergh, J. C. (2010). Externality or sustainability economics? *Ecological Economics, 69*(11), 2047–2052.

Van den Burg, S., Stuiver, M., Norrman, J., Garção, R., Söderqvist, T., Röckmann, C., ... & De Bel, M. (2016). Participatory design of multi-use platforms at sea. *Sustainability, 8*(2), 127.

Victor, P. A. (1991). Indicators of sustainable development: Some lessons from capital theory. *Ecological Economics, 4*(3), 191–213.

World Commission on Environment and Development (WCED). (1987). *Our Common Future.* Oxford: Oxford University Press.

World Economic Forum, *The Circular Economy Imperative,* accessible via https://www.weforum.org/about/circular-economy-videos. Last accessed on 25 September 2019.

Xue, B., Chen, X. P., Geng, Y., Guo, X. J., Lu, C. P., Zhang, Z. L., & Lu, C. Y. (2010). Survey of officials' awareness on circular economy development in China: Based on municipal and county level. *Resources, Conservation and Recycling, 54*(12), 1296–1302.

Zagonari, F., Tsani, S., Mavrikis, S., & Koundouri, P. (2018). Common environment policies in different sustainability paradigms: Evidence from the Baltic, Adriatic, and Black Seas. *Frontiers in Marine Science, 5,* 216.

2 Sustainable Development in Higher Education
A Constructivist Conceptual Framework for Smart Learning and Education

Foteini Paraskeva, Vasiliki Karampa, and Vasiliki Brinia
University of Piraeus and Athens University of Economics and Business

CONTENTS

2.1 INTRODUCTION

According to the United Nations (UN) Sustainable Development Goal (SDG) 4, target 7, every individual is entitled to a good quality education (UNESCO and Sustainable Development Goals, 2019). The quality stems implicitly from the education and culture of people and the empowerment of collective thinking about a sustainable future, and vice versa. Sustainable development leads to high-level services and gifted citizens with augmented skills and can, therefore, strengthen the quality of education (Karampa and Paraskeva, 2019). In our digital era, where technological achievements can be experienced in everyday life's sectors, there is a need for education to be focused on the development of soft, career, and life skills: ICT (Information and Communication Technology) literacy, cross-cultural knowledge and understanding, digital competencies and innovative thinking. In this way, students' lives, learning, and survival in complex and diverse workplaces can be achieved. At the same time, effective and efficient solutions to the problems of the mankind can be further produced and maximized (Luna Scott, 2015a). In that sense, Education for Sustainable Development (ESD) is vital.

Considering that sustainability and future well-being are not a simple matter and that digital technologies master the ways of communication, thinking, feelings, and humans' social interaction, more emphasis should be placed on the enhancement of knowledge and learning based on well-instrumented frameworks, relevant to social-constructivism pedagogical approaches. In this context, innovative teaching methods and learning strategies leverage the learning process by deploying learner-centered practices, composing smart pedagogies. Research highlights a variety of such pedagogies while encompassing master methodologies, for example, collaborative learning, active learning, STEM- and game-based learning, and gamification, among others (Uskov et al., 2019). Combining them properly following fundamental learning principles/guidelines, we are led to the orchestration of robust educational solutions, which can, in turn, be integrated into curriculums related to ESD.

Within the context of active learning, the project-based learning (PjBL) framework is founded on inquiry, collaboration, and reflection. Similarly, STEM is integrated into the instructional design within purely e-learning virtual laboratories or even better in blended authentic environments like flipped classrooms. The learning process can be bolstered, as STEM follows the 5E's retrospective phases of a continuous teaching, learning, and assessment cycle of engagement, exploration, explanation, elaboration, and evaluation through its epistemology and interdisciplinarity; along these lines, opportunities for high-order thinking skills deployment and ICT literacy are cultivated for both teachers and learners (Bybee, 2015). Innovative pedagogical approaches branch off from active/inquiry-based learning such as learning by doing or learning by design, correspondingly trigger innovation, and make creative vision further. On the other hand, and with no doubt, games in learning foster engagement, self-motivation, and better performance and reflect competitiveness and cooperativeness in real entrepreneurial environments. In addition, gamification is evaluated through psychological and motivational learning factors, such as self-regulation and self-direction, self-awareness and self-efficacy, while the incorporation of collaborative strategies and techniques in diverse contexts cultivates the great potential of transcultural understanding and cooperation, empathy, compassion, and solidarity. In particular, when the stimuli concern environmental or ethical issues like climate change or immigration and cultural diversity, more attention should be paid to parameters for the accomplishment of learners' global awareness and consciousness.

Therefore, emphasis is placed both on higher education and undergraduates/postgraduates who directly step forward to career paths and on universities who constitute the educational stakeholders that can bridge the gap between education and the professional world through an ESD.

To this end, this chapter is organized in sections that contain the literature review, the methodology, the results and discussion, and the concluding remarks, trying to answer the following fundamental questions:

- What is the role of education in achieving sustainability?
- What is the importance of tertiary ESD?
- How can SD be integrated into the academic curriculum?
- Why is teaching sustainability using an active learning constructivist approach the best approach to promoting and establishing sustainable practices in the world?
- What is the role of the educational scenarios integrating flexible frameworks for a strategic and a bold vision of the SD in the future?

2.2 LITERATURE REVIEW

This section contains the literature review regarding the role of ESD in higher education, competencies, smart pedagogies, and smart learning based on constructivist approaches and dimensions, as well.

2.2.1 The Vision for Sustainable Development and the Role of Education

In September 2015, a novel set of 17 SDGs was adopted by the UN, based on the already proposed Millennium Development Goals (MDGs), as reference goals for the international community for the period 2015–2030 (UNESCO and Sustainable Development Goals, 2019). These goals were distributed into a network of interconnected targets in an effort of combining economic, social, and environmental areas and where education belongs to as a core component (Le Blanc, 2015). That's because not only education is a fundamental right for humans, but at the same time, it improves them by offering cognitive and noncognitive skills, values, and attitudes in order to live together in harmony and contribute to the well-being, multidimensional sustainability, and evolution of the mankind.

SDG 4 (UNESCO and Sustainable Development Goals, 2019) is committed to the provision of an inclusive and equitable, good quality education to all world citizens throughout their lives, by 2030, since this declaration could work conversely. Citizens who have already been gifted with the privilege of a high-quality education possess a range of strong competencies capable of maintaining this quality at high levels. Particularly, target 4.4 (UNESCO and Sustainable Development Goals, 2019) indirectly confirms the abovementioned assumption, by substantially promoting the increase of youth and adults who have relevant skills, including technical and vocational skills, for employment, decent work, and entrepreneurship. From one point of view, this means that distinct entities such as educators and learners who are integral parts of any educational ecosystem should be provided with a variety of lifelong learning opportunities in order to meet global needs, by using a wide range of education and training modalities. These learning opportunities should broadly focus on the enhancement of problem-solving, critical thinking, creativity, teamwork, communication skills, and conflict resolution, which can be used across a range of occupational fields, beyond the work-specific skills. More specifically, learners need to be prepared for entrepreneurship and life, in accordance with the fast-changing demands of the labor markets, unemployment, and rapidly accelerating technological achievements. In other words, learners need to be trained to acquire the capacity of effective communication and collaboration with others (P21, 2007) from different social or cultural backgrounds, in all learning and working circumstances, and reflect this by participating in society productively (Redecker et al., 2011). They need to cultivate accountability and responsibility as well as adaptivity and flexibility to market changes, while at the same time leadership and initiative are key features to professionalism (Barry, 2012). Additionally, in order to feel competent using technology effectively towards the global trends of "digitalization" and "internetization," namely, the ICT penetration in everyday sectors, they need to develop digital competencies—framed by the European Commission into DigComp (2019)—such as information and data literacy, digital content creation, and problem-solving as well.

Similarly, educators need to be re-skilled or up-skilled, as to be more qualified (target 4.c, UNESCO and Sustainable Development Goals, 2019). Every educator needs to keep up with emerging technologies and smart pedagogies in order to respond to the transformed new role in the digital era (Khlaif and Farid, 2018) and, hence, become "...the guide, facilitator and mediator of learning; the teacher as a learner" (Gros, 2016). In addition, every educator ought to possess a set of professional, emotional competencies, affective skills, aspects, social-cultural, and interpersonal characteristics, mastery of teaching and learning contents as well as pedagogical skills (Huda et al., 2016).

From another point of view, all educational stakeholders are responsible for promoting sustainable development through educational and training programs. This is precisely highlighted in target 4.7 (UNESCO and Sustainable Development Goals, 2019), which articulates the need for acquired cognitive and noncognitive skills by all learners in order to boost sustainable lifestyles, gender equality, promotion of a culture of peace, human rights and nonviolence, appreciation of cultural diversity as well as global citizenship and international understanding. However, emphasis should be placed on higher education, as universities are not only students' forerunners in their life, but also career advances and future challenges in general (David, 2004). In fact, they promote scientific knowledge and research, prompting policy makers to establish regulations in alignment with SDGs (Biermann et al., 2017). Their pivotal and influential role in society in achieving a sustainable future (Cortese, 2003) is further discussed in the section to follow.

2.2.2 EDUCATION FOR SUSTAINABLE DEVELOPMENT IN HIGHER EDUCATION

What does sustainability exactly mean? Considering the definition, Kates et al. (2001) as well as Wiek et al. (2012) provide sustainability "...is the collective willingness and ability of a society to reach or maintain its viability, vitality, and integrity over long periods of time, while allowing other societies to reach or maintain their own viability, vitality, and integrity" (p. 241). Needless to say that the promotion of sustainability is not a simple issue at all. It implies collective intelligence and well-organized endeavor from the society as a whole rather than from distinct, isolated individuals. This responsibility lies on educational institutions, which are a microcosm of the larger society. As a matter of fact, attention is centered on universities, the society's moral supporters bearing in mind they are deeply responsible for learners' awareness, knowledge, skills, and values for a sustainable future. Concurrently, their contribution to instilling the desired values and behaviors in the whole community should not be ignored.

To enhance sustainability, higher education institutions' paradigm should be transformed towards a systemic perspective (Cortese, 2003). This means that the context of higher education, containing highly specialized areas in traditional disciplines, needs transformation into a new, more informal, student-centered (Cörvers et al., 2016); collaborative; and above all, interdisciplinary and transdisciplinary (Brown et al., 2010; Brundiers et al., 2010; Remington-Doucette et al., 2013; Wiek et al., 2015) educational paradigm. By extension, this implies curricula expansion or reform and pedagogical support, by creating and integrating academic programs for sustainable development (ESD), focused entirely on sustainability. At the same time, it requires the enhancement of international collaborations, engaging students in communities and intercultural societies. Moreover, it demands new initiatives for a holistic approach design based on a framework, highlighting graduates' competencies to make decisions considering the long-term future of the economy, ecology, and equity of all communities (UNESCO, 2008).

The systematical integration of ESD programs into their scope of Community engagement, Operations, Research, and Education, the so-called CORE activities (Jenssen, 2012), has occupied many research areas concerning knowledge and competencies students will gain after training. Many researchers get engaged with the key competencies for sustainable education providing frameworks (Wiek et al., 2011), while others investigate best pedagogical practices and carry out competencies' assessments (Remington-Doucette et al., 2013) for specific learning outcomes in order to design and integrate in academic programs successfully. Furthermore, there is a lot of discussion in international literature regarding the effective integration of sustainability in traditional interdisciplinary courses (Remington and Owens, 2009) or the creation of entire sustainability courses and department majors in sustainability.

Our views focus on the pedagogical perspectives, namely, the context of teaching and learning, so as to provide effective solutions for education for sustainability. Hence, to this end, there is a demand for adopting smart pedagogies, founded on applying and reflecting thinking into action from the basic pedagogical theories based on social constructivism to the applied ones such as PjBL; competency-based frameworks; active, experiential, and inquiry-based learning; and actual real-world problem-solving as well. These approaches can enhance cognitive, socio-emotional, and behavioral learning outcomes by promoting an interdisciplinary approach to ensure competency-based education, including skills for entrepreneurship competence frameworks through education and training, promoting a culture in a globalized world. All these issues are discussed in the following paragraphs.

2.2.3 Competencies for Sustainability

As mentioned already, competencies are the key target of today's education in general and education for sustainability. Academia and scholars have proposed frameworks related to competencies and skills, depending on the context of learning, individuals within this context as well as the objectives of the framework. In pursuit of sustainable development, universities primarily need to prepare learners in order to become sustainability professionals (Cörvers et al., 2016). A wide array of research has conducted towards the key competence (i.e., Cusick, 2008; Kelly, 2006; Kevany et al., 2007; Segalas et al., 2009), generally conceptualized as "having the skills, competencies, and knowledge to enact changes in economic, ecological, and social behavior without such changes always being merely a reaction to pre-existing problems" (de Haan, 2006, p. 22). Research results include proposals; however, they present some isolated competencies, rather than a coherent framework of sustainability research and problem-solving competence. After a rigorous literature review across this research discipline in higher education, Wiek et al. (2011) proposed an integrated framework, which encompasses a set of five key competencies in sustainability education, namely, (a) systems thinking competence, (b) anticipatory competence, (c) normative competence, (d) strategic competence, and (f) interpersonal competence. A definition and a brief description of each one of these competencies are presented:

1. Systems thinking competence is also reported by scholars as systemic thinking, interconnected thinking, or holistic thinking (e.g., Clayton and Radcliffe, 1996) and refers to the capacity of collectively analyzing complex systems, namely, identifying their structure, components, and dynamics in technological, economic, and environmental sectors and at a local or global scale. This implies the examination of overlapping effects, inertia, feedback loops, and other systematic features related to sustainability issues and frameworks for solving sustainability problems.
2. Anticipatory competence is also synonym to the terms of future thinking, foresighted thinking, and transgenerational thinking (e.g., Kelly, 2006). It encompasses the systems thinking competence considering that it involves the ability to collectively comprehend and articulate the structure, the key components, and the dynamics of future aspects, including concepts such as time and uncertainty, related to sustainability issues and sustainability problem-solving frameworks. However, analysis is an aid for the evaluation and creation of new methods and methodologies for addressing key issues of sustainability.
3. Normative competence is the ability based on the gained normative knowledge, which includes concepts of justice, equity, social-ecological integrity, and ethics, and refers to collectively mapping and identifying undesirable states and dynamics as well as envisioning desirable ones. At the same time, it encompasses the ability for implementation; reconciliation; and negotiation of values, principles, and goals for sustainability. To this end, there is a need for the abilities necessary for bringing about deliberative change.

4. Strategic competence is linked to the three abovementioned competencies because it mainly incorporates strategies for intervention and transformative change as follows: The current state of the social-ecological system is identified through systems thinking. Sustainable states and dynamics are identified by the normative competence support, which considers existing path dependencies and which might lead to undesirable future states identified by the anticipatory competence. More specifically, strategic competence refers to the ability to collectively developing transformational governance interventions, transitions and strategies for sustainability, and understanding of strategic concepts such as systemic inertia, path dependencies, carriers, and alliances. In international literature, it is also noted as action-oriented competence, transformative competence, and implementation skills.

5. Interpersonal competence is closely related to the abilities of communication, collaboration, leadership, transcultural thinking, and empathy. It is broadly known as collaborative; participatory; interdisciplinary; civic competence to motivate, enable, and facilitate collaborative and participatory sustainability research; and problem-solving.

In addition, the authors suggest a meta-competence, as a critical sixth competence, which refers to the meaningfully use and integration of the five key competencies for sustainable development. The organization of such competency-based frameworks is vital for sustainability; however, smart pedagogies are still the background for the successful application and the potential positive outcomes.

2.2.4   SMART PEDAGOGIES AND SMART LEARNING
FOR SUSTAINABLE DEVELOPMENT

Pedagogy is the nucleus of the rotating concepts of teaching and learning. Pedagogy provides methods and practices of how teaching could foster learning. In the past, traditional pedagogy mainly aimed at cognitive acquisition. Conversely, today's pedagogy has no longer the same focus because students learn in a different way. They are themselves, the center of pedagogy and learning. They are active learners rather than spectators (Luna Scott, 2015b). Today's students belong to the new kind of generation so-called millennials and generation Z (Howe and Strauss, 2000), or digital natives (Prensky, 2001). This digital pervasiveness has led to the transformation of learning environments, not only regarding the penetration of digital accomplishments in different learning settings, but also concerning the effective didactics and the competencies acquirement within them. This kind of environment that can potentially provide optimized services is indicated by researchers and academia (Huang et al., 2012; Hwang, 2014; Koper, 2014; Merrill, 2013; Mikulecky, 2012; Zhu and He, 2012; Zhu et al., 2016; Spector, 2014, 2016, etc.) as smart learning environments (SLEs). They are typically "physical" environments (ASLERD, 2016), that is, a composition of physical and digital components (i.e., smart digital technologies into physical workspaces (Koper, 2014), either physical objects into virtual settings such as augmented reality applications, or other mixed forms). From another perspective, SLEs are mainly a blend of smart pedagogy and smart technology with a variety of affordances for the world sustainable development (Karampa and Paraskeva, 2019). According to Zhu et al. (2016), SLEs are drawn upon smart pedagogies, while instructional strategies and practices take place in both the physical and the digital world, in formal and informal settings. These strategies provide learners class-based differentiated instruction, namely, teaching and learning tailored to learners' preferences; group-based collaborative learning that exploits computer-supported collaborative learning (CSCL) scripts; individual-based personalized learning, which fosters self-regulation and engagement; and mass-based generative learning for meaningful knowledge and enhancement of metacognitive abilities.

Therefore, a great emphasis should be placed on the ingredient of smart pedagogy, since it reshapes learning by offering the smart dimension, and thereby, it contributes to the cultivation of smart learners, consequently future "problem solvers," "change agents," and "transition managers" (Rowe, 2007; McArthur and Sachs, 2009; Willard et al., 2010). There are many statements and conceptualizations in the international literature regarding the terms of "smart pedagogy" and "smart learning." Smart pedagogy is the cause, while smart learning is the outcome of the educational process. Recently, Uskov et al. (2019 p. 2) defined smart pedagogy (SmP) as "...a set of instructor's teaching strategies, activities, and judgements to a) understand the student/learner profile (background, goals, skills, competencies, and capabilities) and b) provide optimal learning processes and environments with corresponding smartness features to help students to achieve their goals." A variation of these words is that smart pedagogy is the lever of innovative teaching methods tailored to current learners' needs and technology incorporation for successful teaching practice and optimal smart learning. Trying to determine smart learning, Gwak (2010) stated that the focus of smart learning is the learners and the content rather than devices; however, it is based on technologies in order to be effective. Kim et al. (2012) similarly noted that smart learning combines the advantages of social learning and ubiquitous leaning; it is learner-centric and a service-oriented educational paradigm. In addition, Scott and Benlamri (2010) and Hwang (2014) argued that smart learning is context-aware and ubiquitous. Lee et al. (2014) proposed that the features of smart learning include formal and informal learning, social and collaborative learning, personalized and situated learning, application and content focus. Middleton (2015) also advocated how smart technologies could contribute to the accomplishment of smart learning and highlighted the learner-centric aspects as well. In general, the present paper echoes the argument of Budhrani et al. (2018) who consider that smart learning is composed of three elements, namely, SLE, pedagogy, and learner. The latter coexists and evolves in all definitions and conceptualizations throughout the years. The conclusion is that smart learning is strongly established in pedagogy, needed when incorporating technology is smart and personalized and lead learners to more independent, informal, and engaging learning paths. It is obvious that definitions as well as core features of smart learning, accomplished by SLEs, are established on pedagogical concepts deep-rooted in social constructivism theory. Since any learning environment is a unique combination of technological, pedagogical, and social components (Kirschner et al., 2004), the following sections discuss these three elements.

2.2.5 THE CONSTRUCTIVIST TECHNOLOGICAL DIMENSION

Constructivist learning constantly takes place in formal settings, such as traditional classrooms (Vintere, 2018). However, the pervasiveness of emerging technologies in schools, universities, and other educational institutions worldwide accentuated the ubiquitous dimension of learning and novel forms of learning environments. Emerging technologies provide usable and user-friendly collaborative tools, fast and readily available communication delivery solutions for effective social interaction through the Internet. Technological developments and computers' mediation to the learning process simultaneously enhanced collaborative learning. Thus, CSCL provides many opportunities for students to engage in group problem-solving activities and paves the way for the consolidation of other key competencies such as metacognition and self-regulation (Hurme and Jarvela, 2001; Lazakidou and Retalis, 2010; Pifarre and Cobos, 2010).

As mentioned in a previous section, new generation's learners have acquired features that justify the term "socio-digital participation" (Hietajärvi et al., 2015), such as flexible use of digital media, multitasking, intellectual ICT tools, socio-digital networking, making and sharing in groups, extended networks, and knowledge creation (Mynbayeva et al., 2017). In addition, emerging technologies enabled various learning communities to be linked together in new and different ways, formulating automatic matched communities and communities of practice, where interaction and communication may occur anywhere and anytime through context awareness. By extension, this means that learning resources are delivered on-demand and will be identified according to the learning event (time, place, peers, and activities). Hence, emerging technologies strongly fostered constructivist learning to be implemented in more feasible ways (Vintere, 2018).

2.2.6 THE CONSTRUCTIVIST PEDAGOGICAL AND SOCIAL DIMENSION

It is widely accepted that the need of humans to support each other when problems appear, has guided them to live and work in various communities (Wang, 2009). Consequently, this fundamental concept guided the theory of constructivism (social constructionism) to be formulated. Social reality has objective and subjective meanings. A social reality is built around every person through the exploitation of the language as a means of communication, knowledge, and understanding. As such, through personal activity, the processes of sociopsychological construction of the society are considered (Mynbayeva et al., 2017).

In education, social constructivism is associated with the socialization skills and the learning of self-structuring knowledge by the students. Social constructivist approach is extended into two perspectives. The first refers to the construction of the learning environment, including communication, and the second includes the construction of knowledge through communication and socialization. According to various pedagogical studies, it is believed that the constructivist approach radically alternates the process of teaching and learning subjects and disciplines, because it connects teaching and learning to daily life. Constructivist ideas in pedagogy have evolved since the 17th century, based on the philosophers of that time, who believed that a person can understand only what he had constructed himself. I. Kant considered that knowledge depends on the interactions between the environment and person's internal properties, and the person himself organizes and interprets his own experience. The idea of constructivism is based on the fact that the human brain does not directly reflect the external world but constructs its experience and life in cognitive and emotional processes in the social context as subjective ideas and concepts. Constructivist ideas are also found in several works, including the ideas of theorists, including Dewey, Piaget, Vygotsky, Candy, Driver, Merizow, Boud. Specifically, the Russian psychologist Vygotsky (1978) emphasizes the importance of the social environment in cognitive development as well as culture and people as the most important factors in the development of an individual. He proved that an individual cannot develop without interacting with the environment. Constructive processes are particularly strong in group conditions, where everyone has a sense of the complex social interactions in which he is included and knowledge starts at the social level, and only then does it become individual knowledge.

The modern methodological principle of addition and complementation justifies the evolution of constructivism theory over the years (Vintere, 2018). Currently, the theory is actualized by using active and collaborative innovative teaching methods (Mynbayeva et al., 2017), divided into nonimitative (i.e., brainstorming, pedagogical exercises, and discussions) and imitative (nongame, e.g., case study, training, etc., and gaming business, e.g., role-playing, blitz games, etc.) in all learning settings and contexts. As such, teachers' role has evolved, in order to become moderator and facilitator, setting up grounds, norms, encouraging participation, monitoring progress, and providing information. More specifically, in the context of higher education and sustainability, the following five constructivism aspects related to the abovementioned ideas further need to be considered more when building the learning process (Bognar et al., 2015):

- Previous knowledge and cognitive structure are the base upon which learners link and construct their new knowledge because learning is the process of interaction between what we know and what we still need to learn;
- Learning occurs when people interact with each other and is, hence, a social process;
- Learning occurs under social and cultural circumstances. In that sense, it is a situational process;
- Learning occurs when people acknowledge and comprehend what skills and strategies are needed for successful outcomes. Therefore, it is a meta-cognitive process;
- Learning is learner-centric and is actually based on the students' activity and autonomy.

To sum up, the pedagogical and social dimension of learning aims to provide and maintain a friendly and interactive environment in which learners feel comfortable and are capable of interacting with each other, as well as active, engaged, and reflective individuals.

2.3 METHODOLOGY

In order to propose a conceptual framework, it is appropriate to first introduce inquiry-based learning models, by giving a greater emphasis on instructional design with PjBL. However, the main goal is to present a three-dimensional, well-orchestrated approach that will incorporate smart pedagogy based on social constructivism, strategies that will enhance the learning process as well as interdisciplinarity of STEAM (Science, Technology, Engineering, Arts, and Mathematics) that will engage the learners and enhance their sustainability key competencies.

2.3.1 PROBLEM AND PROJECT-BASED LEARNING MODELS, INQUIRY, AND STEM

Many researchers and scholars (Cörvers et al., 2016) advocate the pedagogical integration of problem-based learning (PrBL) and PjBL through real-world sustainability issues, because these two models or a hybrid form could offer opportunities for the enhancement of learners' sustainability competencies. More specifically, Cörvers et al. (2016) mention that these two educational models incorporate the four key principles, namely, constructive learning, collaborative learning, contextual learning, and self-directed learning. Accordingly, learners acquire new knowledge by constructing this knowledge based on the already existing one, interacting with peers in various groups, transferring knowledge through real-world problems, and undertaking the regulation of their own learning process by themselves.

In particular, the hybrid form of these two models combines both their advantages, offering added value to sustainability education. According to Brundiers and Wiek (2013), the hybrid form illustrates students' engagement in real-world tasks and stimuli for professional situation, gives a new dimension to teacher's role as facilitator and resource guide, actuates students to the process of multiple information sources, and establishes conditions for formative, self, and peer-to-peer assessments.

Nevertheless, the present study is oriented towards the PjBL model, because it engages learners in continuous, collaborative research (Bransford and Stein, 1993). Based on a sequential procedure enriched by a variety of instructional strategies and methods, it provides practical products as learning outcomes and applicable results (Brundiers and Wiek, 2013). The PjBL model emphasizes the process of constructing artifacts through complex problem-solving and decision-making activities based on challenging questions. Additionally, it reflects constructivism by the development of a learner-centered learning environment, which includes authentic content, authentic assessment, facilitating teaching but not direction, clear educational goals, collaborative learning, and reflection (Thomas, 2000), among others. Hence, learners are able to work relatively autonomously over extended periods of time by enhancing time management skills and create knowledge actively by exploiting their authentic and real-life experiences.

The PjBL projects are focused on activities during which learners are faced with central concepts and principles of a discipline (Thomas, 2000). However, PjBL springs off inquiry-based learning and encourages STEM interdisciplinary methodology. According to research findings, inquiry-based learning (Kalsoom and Khanam, 2017) and mutual learning implicit in the STEM interdisciplinary approaches (Clark and Button, 2011) enhance the sustainability comprehension of students in higher education. In specific, Kalsoom and Khanam (2017) conclude that when art and aesthetics converge with science, students have many possibilities to strongly understand eco-justice for the promotion of a sustainable society, and thus, STEM is implicitly translated to STEAM.

2.3.2 PROPOSED CONCEPTUAL FRAMEWORK FOR ESD

The issues discussed before, in all previous sections of this chapter, have motivated the current study to proceed to a wider research proposition of a constructivist conceptual framework in order to support smart learning and ESD in higher education. The proposed conceptual framework is clearly imprinted in Figure 2.1.

Obviously, the proposed framework is composed of a set of different fundamental components and works as a flexible platform for educators and learners. These components contain input and output parameters as well as the core SLE, which is visualized as a 3D dynamic structure. Primary and external data basically related to learners' characteristics as well as additional case studies and real-world problems are provided as input to the SLE. This, in turn, functions as an integrated processing system, partially abstract represented by three sides of a three-dimensional (3D) structure. The first side of dimension refers to an essential repository, filled with all the incoming information. The second side of dimension is represented by the educational background of STEAM disciplines. STEAM could be adapted to a variety of incoming real-world problems structuring a coherent interdisciplinary

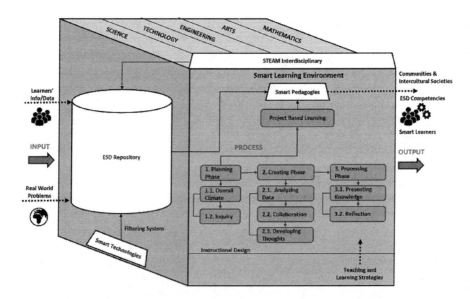

FIGURE 2.1   A three-dimensional proposed framework for ESD.

content. Considering learners' needs and choices as well as additional incoming data, a smart filtering technological system encodes the preferences and provides content delivery. The third side is related to the pedagogical dimension. Smart pedagogies, based on social constructivist theories, encompass the model of PjBL, enriched by instructional methods, and teaching and learning strategies. The aim and the output of the process are the cultivation of sustainability key competencies, the production of effective sustainability solutions to communities and intercultural societies, and finally the upbringing of the new generation's learners, namely, the smart learners. A short analysis of the proposed conceptual framework is presented below:

- **Input:** It contains all the incoming external data, mainly referring to learners' characteristics and needs, such as demographic information (age, sex, etc.) and educational preferences (face-to-face, e-learning, blended settings). At the same time, real-world problems, particularly related to learners' location and social background, are provided as input to the processing procedure of the SLE.
- **Process:** Through the SLE, real-world problems become the vehicle for sustainability research and problem-solving key competencies' cultivation. In particular, this is achieved by integrating the STEAM interdisciplinary approach adapted to the real-world inbound case studies loaded into the ESD repository along with all necessary user data. Adaptive and context-aware smart technologies elaborate these data by using predetermined criteria and produce a variety of learning content. On the other hand, educators design learning courses by exploiting the processed information and the delivered content. Hence, they are able to set a number of learning goals and follow the instructional design provided by the PjBL, potentially enhanced by a variety of instructional strategies and methods. Particularly, the present study proposes the PjBL model as it is adopted by Han and Bhattacharya (2010) into the guide "Emerging Perspectives on Learning, Teaching, and Technology" of the Department of Educational Psychology and Technology, at the University of Georgia. The model includes three basic phases and six individual subphases, accompanied by a short description and potential strategies, and is presented in Table 2.1.
- **Output:** Sustainability-effective solutions and smart learners' generation augmented with skills and sustainability key competencies empower societies and promote sustainable development.

2.4 RESULTS AND DISCUSSION

Higher education institutions hold a crucial role in the context of ESD because they are mainly graduates' precursors to their professional development as well as academic contributors to the promotion of research and the development of sustainability. The integration and consolidation of academic programs for sustainable development is vital towards the emergence of learners' skills, knowledge, and value-based development, in order to be active citizens in creating a more sustainable society (DfES, 2005). PjBL is a pedagogy that involves students in higher education in

TABLE 2.1

The PjBL Model (Han and Bhattacharya, 2010)

Phases	Subphase	Description	Strategies
1. Planning phase	1.1. Designing overall climate	Development of an environment that will promote inquiry, challenges, and collaborative research	• Investigate challenging problems • Create provoking questions
	1.2. Inquiry	• Themes and topics selection • Resource sharing • Organization of collaboration	• Focus on concepts and skills central to disciplines • Commit themselves to completing work in groups • Engage in work • Use the tools
2. Creating phase	2.1. Analyzing data	• Deployment of cases • Research design • Perform an experiment • Data collection and analysis	• Learn how to work • Learn to use project management processes
	2.2. Collaborating with others	Collaboration and communication in order to find out solution	Work in teams to complete complex tasks
	2.3. Developing thoughts and documentation	• Construction of the artifacts • Ideas construction and visualization	• Use the processes of design thinking • Manage a multistep project
3. Processing phase	3.1. Presenting knowledge and artifacts	• Presentation of knowledge • Application of the proposed solution	• Share their work-in-progress with peers and teachers • Receive feedback
	3.2. Reflection and follow-up	• Self-assessment • Peer-to-peer assessment	• Reflect on concepts, process, and learned skills • Learn to assess and suggest improvements in their own and other students' work

Source: Han and Bhattacharya (2010).

applying and developing theories, skills, and techniques to solve real-world problems by participating in authentic communities. PjBL reflects constructivism because it incorporates inquiry-based, active, collaborative, and self-directed learning, namely, places learners into the center of learning.

On the other hand, it is widely acknowledged that learning has been enhanced and augmented by the diffusion of digital technologies. The digital pervasiveness transformed learning environments into novel and SLEs, where smart technologies support at the fullest the learning process. As such, adaptive and context-aware smart technologies, in combination with smart pedagogies, namely, incorporated innovative methods and strategies as well as interdisciplinary STEAM methodology integration, provide opportunities to foster key sustainability competencies to higher education students and novel and effective solutions for world sustainable development.

2.5 CONCLUDING REMARKS

This chapter deals with the integration of education of sustainable development programs in higher education, in order to enhance learners' core competencies, to provide effective solutions and promote sustainability to the global society. Therefore, after a literature review, a conceptual framework is proposed, established on the essential concepts of SLEs, those of smart technologies and smart pedagogies grounded on social constructivism learning theory, particularly taking into account the STEAM disciplines. However, the conceptual framework is theoretically developed, and thus, research evidence is needed for its documentation.

REFERENCES

ASLERD. (2016). Timisoara Declaration. Better learning for a better world through people centred smart learning ecosystems. Retrieved September 19 from: www.mifav. uniroma2.it/inevent/events/aslerd/docs/TIMISOARA_DECLARATION_F.pdf.
Barry, M. (2012). *What Skills Will You Need to Succeed in the Future?* Phoenix Forward. Tempe, AZ: University of Phoenix.
Biermann, F., Kanie, N., & Kim, R. E. (2017). Global governance by goal-setting: The novel approach of the UN Sustainable Development Goals. *Current Opinion in Environmental Sustainability, 26*, 26–31.
Bognar, B., Gajger, V., & Ivic, V. (2015). Constructivist e-learning in higher education. *Croatian Journal of Education: Hrvatski časopis za odgoj i obrazovanje, 18*, 31–46.
Bransford, J. D. & Stein, B. S. (1993). The IDEAL problem solver.
Brown, V. A., Harris, J. A., & Russell, J. Y. (Eds.) (2010). *Tackling Wicked Problems through the Transdisciplinary Imagination*. London: Earthscan.
Brundiers, K., Wiek, A., & Redman, C. L. (2010). Real-world learning opportunities in sustainability: From classroom into the real world. *International Journal of Sustainability in Higher Education, 11*(4), 308–324.
Brundiers, K. & Wiek, A. (2013). Do we teach what we preach? An international comparison of problem-and project-based learning courses in sustainability. *Sustainability, 5*(4), 1725–1746.
Budhrani, K., Ji, Y., & Lim, J. H. (2018). Unpacking conceptual elements of smart learning in the Korean scholarly discourse. *Smart Learning Environments, 5*(1), 23.
Bybee, R. W. (2015). *The BSCS 5E Instructional Model: Creating Teachable Moments*. Arlington, VA: NSTA Press, National Science Teachers Association.
Clark, B. & Button, C. (2011). Sustainability transdisciplinary education model: Interface of arts, science, and community (STEM). *International Journal of Sustainability in Higher Education, 12*(1), 41–54.
Clayton, A. M. & Radcliffe, N. J. (1996). *Sustainability: A Systems Approach*. Boulder, CO: Westview Press.
Cortese, A. D. (2003). The critical role of higher education in creating a sustainable future. *Planning for Higher Education, 31*(3), 15–22.

Cörvers, R., Wiek, A., de Kraker, J., Lang, D. J., & Martens, P. (2016). Problem-based and project-based learning for sustainable development. In: H. Heinrichs, P. Martens, G. Michelsen, & A. Wiek (Eds.) *Sustainability Science* (pp. 349–358). Dordrecht: Springer.

Cusick, J. (2008). Operationalizing sustainability education at the University of Hawai 'i at Manoa. *International Journal of Sustainability in Higher Education, 9*(3), 246–256.

De Haan, G. (2006). The BLK '21' programme in Germany: A 'Gestaltungskompetenz'-based model for Education for Sustainable Development. *Environmental Education Research, 12*(1), 19–32.

Department for Education and skills (DfES). (2005). *Developing a Global Dimension to the School Curriculum* (rev. ed.). London: DFES DFID DEA.

European Commission. (2019). The digital competence framework 2.0. Retrieved September 2019 from: https://ec.europa.eu/jrc/en/digcomp/digital-competence-framework.

Gros, B. (2016). The dialogue between emerging pedagogies and emerging technologies. In: B. Gros, Kinshuk, & M. Maina (Eds.) *The Future of Ubiquitous Learning* (pp. 3–23). Berlin, Heidelberg: Springer.

Gwak, D. (2010). *The Meaning and Predict of Smart Learning.* Smart Learning Korea Proceeding, Korean e-Learning Industry Association.

Han, S. & Bhattacharya, K. (2010). Constructionism, learning by design, and project based learning. In: M. Orey (Ed.) *Emerging Perspectives on Learning, Teaching, and Technology*, pp. 127–141. Global Text Project, Zurich, Switzerland.

Hietajärvi, L., Tuominen-Soini, H., Hakkarainen, K., Salmela-Aro, K., & Lonka, K. (2015). Is student motivation related to socio-digital participation? *Procedia: Social and Behavioral Sciences, 171*, 1156–1167.

Howe, N. & Strauss, W. (2000). *Millennials Rising: The Next Great Generation.* New York: Vintage.

Huang, R., Yang, J., & Hu, Y. (2012). From digital to smart: The evolution and trends of learning environment. *Open Education Research, 1*(1), 75–84.

Huda, M., Anshari, M., Almunawar, M. N., Shahrill, M., Tan, A., Jaidin, J. H., … & Masri, M. (2016). Innovative teaching in higher education: the big data approach. *TOJET.* Special Issue for INTE 2016. 1: 1210–1216.

Hurme, T. R. & Järvelä, S. (2001). Metacognitive processes in problem solving with CSCL in mathematics. In: P. Dillenbourg, A. Eurelings, & K. Hakkarainen (Eds.) *European Perspectives on Computer-Supported Collaborative Learning* (pp. 301–307). Maastricht: University of Maastricht.

Hwang, G. J. (2014). Definition, framework and research issues of smart learning environments--a context-aware ubiquitous learning perspective. *Smart Learning Environments, 1*(1), 4.

Jenssen, S. (2012). Sustainability at universities: An explorative research on assessment methods and tools for sustainability implementation at universities. Master of sustainability science and policy, Maastricht University.

Kalsoom, Q. & Khanam, A. (2017). Inquiry into sustainability issues by preservice teachers: A pedagogy to enhance sustainability consciousness. *Journal of Cleaner Production, 164*, 1301–1311.

Karampa, V. & Paraskeva, F. (2019). Smart learning environments: A blend of ICT achievements and smart pedagogy for the World Sustainable Development. *In* International Conference on Human Interaction and Emerging Technologies (pp. 482–488). Cham: Springer.

Kates, R. W., Clark, W. C., Corell, R., Hall, J. M., Jaeger, C. C., Lowe, I., … & Faucheux, S. (2001). Sustainability science. *Science, 292*(5517), 641–642.

Kelly, P. (2006). Letter from the oasis: Helping engineering students to become sustainability professionals. *Futures, 38*(6), 696–707.

Kevany, K., Huisingh, D., García, F. J. L., & Kevany, K. D. (2007). Building the requisite capacity for stewardship and sustainable development. *International Journal of Sustainability in Higher Education, 8*(2), 107–122.

Khlaif, Z. N. & Farid, S. (2018). Transforming learning for the smart learning paradigm: Lessons learned from the Palestinian initiative. *Smart Learning Environments*, 5(1), 12.

Kim, T., Cho, J. Y., & Lee, B. G. (2012). Evolution to smart learning in public education: A case study of Korean public education. *In IFIP WG 3.4 International Conference on Open and Social Technologies for Networked Learning* (pp. 170–178). Berlin, Heidelberg: Springer.

Kirschner, P., Strijbos, J. W., Kreijns, K., & Beers, P. J. (2004). Designing electronic collaborative learning environments. *Educational Technology Research and Development*, 52(3), 47.

Koper, R. (2014). Conditions for effective smart learning environments. *Smart Learning Environments*, 1(1), 5.

Lazakidou, G. & Retalis, S. (2010). Using computer supported collaborative learning strategies for helping students acquire self-regulated problem-solving skills in mathematics. *Computers and Education*, 54(1), 3–13.

Le Blanc, D. (2015). Towards integration at last? The sustainable development goals as a network of targets. *Sustainable Development*, 23(3), 176–187.

Lee, J., Zo, H., & Lee, H. (2014). Smart learning adoption in employees and HRD managers. *British Journal of Educational Technology*, 45(6), 1082–1096.

Luna Scott, C. (2015a). The futures of learning 2: What kind of learning for the 21st century?

Luna Scott, C. (2015b). The futures of learning 3: What kind of pedagogies for the 21st century?

McArthur, J. W. & Sachs, J. (2009). Needed: A new generation of problem solvers. *Chronicle of Higher Education*, 55(40), 1–4.

Merrill, M. D. (2013). *First Principles of Instruction: Identifying and Designing Effective, Efficient and Engaging Instruction*. Hoboken, NJ: Pfeiffer.

Middleton, A. (Ed.) (2015). *Smart Learning: Teaching and Learning with Smartphones and Tablets in Post-Compulsory Education*. Sheffield: Media-Enhanced Learning Special Interest Group and Sheffield Hallam University Press.

Mikulecký, P. (2012). Smart environments for smart learning. *In 9th International Scientific Conference on Distance Learning in Applied Informatics*, Sturovo, Slovakia (pp. 213–222).

Mynbayeva, A., Sadvakassova, Z., & Akshalova, B. (2017). Pedagogy of the twenty-first century: Innovative teaching methods. In Olga Bernad-Cavero (Eds): *New Pedagogical Challenges in the 21st Century-Contributions of Research in Education*. IntechOpen, London, UK.

Partnership for 21st Century Skills. (2007). Retrieved September 2019 from: www.battelleforkids.org/networks/p21.

Peterson, A., Dumont, H., Lafuente, M., & Law, N. (2018). Understanding innovative pedagogies: Key themes to analyse new approaches to teaching and learning. *OECD Education Working Papers*, 172, 1–134.

Pifarre, M. & Cobos, R. (2010). Promoting metacognitive skills through peer scaffolding in a CSCL environment. *International Journal of Computer-Supported Collaborative Learning*, 5(2), 237–253.

Prensky, M. (2001). Digital natives, digital immigrants part 1. *On the Horizon*, 9(5), 1–6.

Redecker, C., Ala-Mutka, K., Leis, M., Leendertse, M., Punie, Y., Gijsbers, G., Kirschner, P., Stoyanov, S., & Hoogveld, B. (2011). *The Future of Learning: Preparing for Change*. Luxembourg: Publications Office of the European Union.

Remington, S. M. & Owens, K. S. (2009). Researching ocean acidification in general chemistry. Curriculum for the Bioregion Classroom Activities for Sustainability, available at the Washington Center for Improving the Quality of Undergraduate Education.

Remington-Doucette, S. M., Hiller Connell, K. Y., Armstrong, C. M., & Musgrove, S. L. (2013). Assessing sustainability education in a transdisciplinary undergraduate course focused on real-world problem solving: A case for disciplinary grounding. *International Journal of Sustainability in Higher Education, 14*(4), 404–433.

Rowe, D. (2007). Education for a sustainable future. *Science, 317*(5836), 323–324.

Scott, K. & Benlamri, R. (2010). Context-aware services for smart learning spaces. *IEEE Transactions on Learning Technologies, 3*(3), 214–227.

Segalàs, J., Ferrer-Balas, D., Svanström, M., Lundqvist, U., & Mulder, K. F. (2009). What has to be learnt for sustainability? A comparison of bachelor engineering education competences at three European universities. *Sustainability Science, 4*(1), 17.

Spector, J. M. (2014). Conceptualizing the emerging field of smart learning environments. *Smart Learning Environments, 1*(1), 2.

Spector, J. M. (2016). Smart learning environments: Concepts and issues. *In Society for Information Technology & Teacher Education International Conference* (pp. 2728–2737). Charleston, SC: Association for the Advancement of Computing in Education (AACE).

Sustainable Development Goals (SDGs), General Assembly. (2015). Transforming our world: The, 2030.

Thomas, J. W. (2000). A review of research on project-based learning.

Vintere, A. (2018). A constructivist approach to the teaching of mathematics to boost competencies needed for sustainable development. *Rural Sustainability Research, 39*(334), 1–7.

Wang, Q. (2009). Designing a web-based constructivist learning environment. *Interactive Learning Environments, 17*(1), 1–13.

Wiek, A., Bernstein, M., Foley, R., Cohen, M., Forrest, N., Kuzdas, C., Kay, B., Withycombe Keeler, L. (2015). Operationalising competencies in higher education for sustainable development. In: M. Barth, G. Michelsen, M. Rieckmann, & I. Thomas (Eds.) *Handbook of Higher Education for Sustainable Development* (pp. 241–260). London: Routledge.

Wiek, A., Ness, B., Brand, F.S., Schweizer-Ries, P. and Farioli, F. (2012). From complex systems analysis to transformational change: A comparative appraisal of sustainability science project. *Sustainability Science, 7*(1), 5–24.

Wiek, A., Withycombe, L., & Redman, C. L. (2011). Key competencies in sustainability: A reference framework for academic program development. *Sustainability Science, 6*(2), 203–218.

Willard, M., Wiedmeyer, C., Warren Flint, R., Weedon, J. S., Woodward, R., Feldman, I., & Edwards, M. (2010). The sustainability professional: 2010 competency survey report. *Environmental Quality Management, 20*(1), 49–83.

UNESCO. (2008). Education for all global monitoring report. Retrieved September 2019 from: http://unesdoc.unesco.org/images/0015/001547/154743e.pdf.

United Nations General Assembly. (2015) Retrieved September 2019 from: https://en.unesco.org/sdgs.

Uskov, V. L., Bakken, J. P., & Aluri, L. (2019). Crowdsourcing-based learning: The effective smart pedagogy for STEM education. *In 2019 IEEE Global Engineering Education Conference (EDUCON)* (pp. 1552–1558). Dubai: IEEE.

Vintere, A. (2018). A constructivist approach to the teaching of mathematics to boost competencies needed for sustainable development. *Rural Sustainability Research, 39* (334), 1–7.

Vygotsky, L. (1978). Interaction between learning and development. *Readings on the Development of Children, 23*(3), 34–41.

Zhu, Z. T. & He, B. (2012). Smart education: New frontier of educational informatization. *E-Education Research, 12*, 1–13.

Zhu, Z. T., Yu, M. H., & Riezebos, P. (2016). A research framework of smart education. *Smart Learning Environments, 3*(1), 4.

3 Cultural Heritage and Education for Sustainability

Vasiliki Brinia
Athens University of Economics and Business

CONTENTS

Whether it's dusk or dawn's first light the jasmin stays always white…

George Seferis
Greek poet and winner of the 1963 Nobel Prize for Literature

3.1 INTRODUCTION

This unit delves into the triangle of concepts of cultural heritage, education, and sustainability, a contemporary matter of thinking forward through thinking back, understanding the old, and building the new in an interaction of past memories and present experiences, which forms the individual's perception of the world. But from this flow of thought-echoing Dewey's opinions, expressed in Art as Experience (1934), are drawn the lines along which to examine the matter and define the relation of the three concepts in hand. This is a descriptive analysis aided by online academic search tools, using a plethora of older studies regarding the concepts and modern-day experiments and educational actions through inter-sectoral collaborations on a practical level (Dewey, 2005).

Thus, it is asked what is the connection between cultural heritage, education, and sustainability and consequently what are the steps to be taken so that the value of cultural heritage can not only be exploited in a viable and sustainable way that produces services which reinforce local economies, but is also enhancing educational results, while the participants of education themselves make value-adding suggestions through project-based learning. In this effort, economic, social, political, and cultural sustainability aspects are taken into account, as well as their connection with education and cultural heritage, separately, and at the same time, in theory and in practice, so that human beings' quality of life is improved, individuals and groups are empowered, social cohesion is enhanced, and economic development is boosted in the long run, with minimal negative externalities and with social inclusion, which even reinvigorates rural communities.

This three-way reciprocity is a recent phenomenon that has not been explored and put into a theoretical context, in spite of some very contemporary practical actions by collaborating institutions, as showcased by specific examples in the way described further on. The whole venture is stimulated by the works of Pines and Gilmore (1998) on the Experience Economy and Agenda 21 (UNESCO, 2002), and on Education, Awareness, and Training, which states that education, formal education, training, and public awareness included is a process through which societies and individuals can reach their full potential, while it is of utmost importance towards public participation in decision-making and the achievement of ethical and environmental awareness, attitudes and values, skills, and behavior in accord with sustainable development. It is also clarified that both formal education and informal education are crucial to modifying people's attitudes so that they have the capacity to assess and address sustainable development issues.

3.2 CONCEPTUAL FRAMEWORK

Beginning with sustainability, before any modern application in management and policymaking, this concept has been used in order to express how in forest management one ought to harvest no more than the growth rate of the forest replenishes (Wiersum, 1995). It is indicative that the German word for sustainability "nachhaltigkeit" is, at first, used in 1731 in that silvicultural context (Wilderer, 2007). Then, in his 1,798 book "An Essay on the Principle of Population," Thomas Malthus (1986) expressed a somewhat pessimistic view on sustainability, comparing the exponential growth of human population compared to the limited resources available.

More modern approaches are that of Harold Hotelling (1931) on optimal rates of use of nonrenewable resources and the 1987 Brundtland Report that defines sustainability as satisfying present needs, while conserving the ability of the next generation to keep meeting them.

The Report itself has a perspective of development and environment, close to the all-time classic economics of needs versus resources or, in this case, present needs versus future needs in accordance with resources.

Lastly, contemporary views of sustainability can be found in the view of sustainable development found in UN Agenda for Development (1997), which views development as a multilayered and multilateral effort towards a better quality of life for all, recognizing sustainable development consisting of economic and social development and protecting the environment.

Thus, it becomes obvious that, today, sustainability is mainly viewed through three perspectives: environmental, economic, and social (Chiu, 2004; European Commission, 2005; Helming, Perez-Soba & Nurse, 2006; Robert, Parris & Leiserowitz, 2005; Strange & Bayley, 2008; Tabbush, 2008; WCED, S.W.S., 1987). There are, of course, suggestions such as the addition of a political and a cultural dimension, proposed by

Robert Gibson (2001), coming from a position of the inadequacy of material goods to measure human prosperity, while Tom Kuhlman and John Farrington (2010) remind that cultural heritage left to the next generations also consists of art and cultural landscapes, as well as infrastructure, technology, and institutions.

According to Randall (2008), capital, in general, could be a substitute for natural resources and thus an answer to the question of sustainability, so cultural capital, in particular, is a pillar of sustainability that produces cultural goods. David Throsby (1999) defines cultural capital as the stock of cultural value embodied in an asset, which produces economic and cultural goods. He categorizes cultural capital into intangible and tangible, with the former being public domain goods—like music and literature, as well as sets of ideas like values, practices, traditions, and beliefs—that help groups bond (see political pillar of sustainability), and the latter including private artistic objects and also what is called cultural heritage in the form of buildings, monuments, sites, and locations of cultural significance.

Another perspective on cultural capital has been proposed by Bourdieu (1986), who distinguishes it into the embodied state pertaining to the cultured individual's mind and body; the objectified state, including cultural goods like pictures, paintings, machines, books, and instruments; and the institutionalized state comprising all kinds of academic credentials and professional qualifications. Bourdieu in his more individualistic inclinations, giving in any case the high ground to the embodied state, does give an angle on cultural capital that approximates that of the concept of human capital in economics (Robbins, 1991), while some economists themselves include culture in human capital, like Costanza and Daly (1992) defining human capital as the stock of culture, skills, knowledge, and education accumulated in human beings.

A connection is then formed between cultural capital and human capital, which is mainly propagated through education. It is, either way, the very idea of leaving something behind for the next generation and putting off present welfare for future prosperity, a very characteristic symbolism for capital with, among others, moral extensions that in the modern society are cultivated through education, as were through religion, with an example being what Max Weber would call the spirit of capitalism, talking of the protestant ethic (2013). The level of dedication to sustainability in education is shown in Chapter 36 Agenda 21 (UNESCO, 2002), and in Education, Awareness, and Training, which states that education, formal education, training, and public awareness included is a process through which societies and individuals can reach their full potential, while it is of utmost importance towards public participation in decision-making and the achievement of ethical and environmental awareness, attitudes and values, skills, and behavior in accord with sustainable development. It is also clarified that both formal education and informal education are crucial to modifying people's attitudes so that they have the capacity to assess and address sustainable development issues.

Thus, education is an important pillar in building the skills, the ethics, and the consensus needed to overcome one's short-term interests and in choosing to take a "leap of faith" and trust one's partners in an inverted prisoner's dilemma (Poundstone, 1992), where sacrificing short-term personal interests leads to more prosperity for one, one's peers, and the future generations. Because there needs to be a reason for one not to tear down the stones of an old building, which could become monumental—if it is not yet—just to drive down the costs of building one's own house or when in a position of authority, to not make a misuse of EU funding towards overpriced projects built around monuments, to boost short-term profits of a friendly contractor and cash-in those benefits regardless of negative externalities, rather than making a viable plan for upgrading services provided and including the community, in the long run.

Agenda 21 Chapter 36 (UNESCO, 2002) does indeed seem on the right direction if one keeps in mind the above, aiming at skills like critical thinking; collaboration and cooperation; oral, written, and graphic communication; creative thinking; problem-solving; evaluation and reflection; planning; decision-making; proper technology; media and ICT (Information and Communication Technology) use; and civil participation and action.

3.3 METHODOLOGY

It is through a descriptive kind of educational research that a historical and content analysis with practical references to qualitative research is constructed. This is the result of processing vast amounts of literature searched for through the HEAL-Link, EBSCO, and Google Scholar search tools, producing an amalgam revolving around a core of concepts, taking a form and a direction that both dictate the literature researched and are, then, influenced by the findings themselves. Thus, the unit busies itself with presenting a context built by centuries of studies on separate concepts and showing how they converge in theory and in practice, through references to practical experiments and educational actions.

3.4 RESULTS AND DISCUSSION

Applied educational efforts on the direction of the conceptual framework can be found in Brinia et al. (2019), where ways to instill cultural values through ancient drama and cultural promenades are explored and the method is evaluated through in-depth interviews after the experience. And it is such a venture that really ties together education, cultural heritage, and sustainability, in practice and through the experiential, as it proposes a teaching methodology using the outdoor and art. More specifically, this field research was conducted by the Teacher Education Program of Athens University of Economics and Business, using 42 teacher-candidates with economics, management, and business administration and IT background in a promenade through Acropolis and then monitoring their perceptions on their own experience with this teaching method, which is described by the authors, step-by-step, as such:

a. The 42 teacher-candidates are separated into seven-member groups. To each team, a coordinator is assigned, who has beforehand planned the tour of the group.
b. The groups are split according to location and play an ice-breaking game, as a group-making activity, guided by the coordinator.
c. A tour is, then, taking place at the Ancient Theater of Dionysus.

    d. Students role-play, using an excerpt from "Wealth" ("Ploutos") by Aristophanes with their arrival in the Theater of Dionysus. Students converse on the deeper connotations of the play and draw connections with the present.

    e. A show-around at the Stoa of Eumenes ensues. Fable of the name of the city of Athens narration is done here. It is discussed how ancient Greek couples had walks in almost the same area as modern couples.

    f. The Odeon of Herodes Atticus is explored, and associations with its more recent history and its value are drawn. Special reference is made by the team coordinator to Maria Callas, who had performed at the Odeon, so as to inspire students through the modern use of the theater, all while a recording of the aria from the opera "Norma," performed by Maria Callas, is playing. Finally, a discussion of the personal experiences of the students, regarding the Odeon, follows.

The learning results of this qualitative research through the in-depth interviews followed showcase the method's effectiveness in instilling cultural values, knowledge, and skills that promote sustainability, are indicated by the answers of the subjects. First of all, the effectiveness of the method finds its first hint in the emotions created, which in this occasion were melancholy or nostalgia for the past, and hope for the future, curiosity, awe, surprise, sadness, pleasure, pride, belonging, and togetherness. Even if it were not for the students themselves that reported feeling, they created lifelong memories and acquaintances and the already-mentioned views of John Dewey and others in the field of experiential education and learning (Kolb, 2014); on learning through experience, emotion, and memory interaction, the thought that emotion and learning are positively connected is prominent in the literature of many fields from neurology and psychology to management and business administration (Weiss, 2000; Antonacopoulou & Gabriel, 2001; Hu et al., 2007).

The students also did get to feel like a part of history, connecting to everyday people of the past and bearing the responsibility for the future, appreciating "today," as well as the value of the monuments visited. They took in new information through role-playing, dialogue, personal contribution, and storytelling and their application on this unique environment, creating multisensory stimuli, in a way that could hardly be achieved through technical means and digital representation, as was also reported by the students. Subjects also expressed respect for their history and felt they should advocate their cultural heritage in their future classes as a form of promoting cultural and economic sustainability, as well as what could be assigned to political sustainability through a mutual feeling of shared identity, reinforced by symbolic thought connecting the patterns of the monuments to their respective eras, then to the landscape, and finally to a people's psyche.

Of utmost importance on the field of morality, which is already stressed as an integral part in a sort of sustainability "software" in human behavior, are the messages emerging from students' answers, due to symbolic thought, their interaction with one another, and the landscape, that had to do with respecting one another, collaboration, resistance to greediness, through the Aristophanes' Ploutos (Wealth) role-play, love, peace, culture, and identity to be sustained and promoted in the long run.

Another interesting technical detail, other than group dynamics and the sensory advantage of visiting the monuments itself, was the use of pauses during the promenade that helped mindfulness, processing stimuli, and reflecting on new information. Such state-of-the art approaches do not materialize out of thin air of course. They stand on a multifield substructure and then contribute back to it, as is seen above, even with a direct connection between the subjects' opinions and older thinkers and researchers. For example, Nassauer (1997) argues for a concept of cultural sustainability including and supporting landscape ecology, while Naveh (2007) proposes a modernized adoption of cultural values towards the goal of developing strategies of sustainability, such that people and their culture are benefited, while also enjoying economic benefits and the urban and rural communities alike are lifted. Tibbs' (2011) views on sustainability also include the change of cultural values as a call for a broader change for the sake of sustainability. The above are not unique in the discussion over cultural sustainability and cultural economy of course (Throsby, 2001, 2008; Hawkes, 2001).

For the above to become reality, such notions must be taught and the way to achieve this finds its foundations as far back as Aristotle's Nicomachean Ethics where it is mentioned that we learn an art or craft by doing the things that we shall have to do when we have learnt it (Wikisource, 2011). Bloom in his Taxonomy of Educational Objectives (1956) lists the stages of learning as follows: knowledge, comprehension, application, analysis, synthesis, and assessment of value, while further works result in the Revised Taxonomy (Cannon & Feinstein, 2014) that presents the cognitive process and the knowledge dimension as the two dimensions of learning. Further elaboration on the experiential learning theory is conducted by Kolb (2014) who is influenced by Dewey, Piaget, and Lewin, and through whom is formulated a four-learning-objective experiential learning cycle, which describes the path from the live experience to a reflective observation of it, ending in abstract conceptualization (Bassi & Russ-Eft, 1997; McLeod, 2017).

Such tools of experiential learning are Outdoor Education and Art Education used in "How to Instill Cultural Values in the New Generation through Cultural Promenades and Ancient Drama" (Brinia et al., 2019) that is analyzed above. First, outdoor education is, according to Simon Priest, an experiential method of learning that takes place in the outdoors, makes use of all senses, best fits in an interdisciplinary curriculum, and involves natural resources as well as human relationships (Priest, 1986). With recreational, cultural, and training uses of promenades through history in mind and specifically as described by Peter Borsay concerning 17th-century England (Borsay, 1986) and by O'Byrne concerning 18th-century London (O'Byrne, 2006), it is fair to say that cultural promenades do in fact constitute a form of outdoor education. In this way, the students are given an opportunity to think over concepts like spirituality, culture, and myth, through contact with their environment (Terkenli, 1996). The very environment, through its interaction with human elements, shapes their societies and the rituals they build their collective and personal identity with (Stavridis, 2006).

Similarly, ancient drama can be used as part of art education (Isyar & Akay, 2017), as it echoes social and cultural values rooted in tradition and cultural heritage and promotes social and cultural diversity from the perspective of Democratic

Education (Harfitt & Chow, 2018; Moss, 2008). Also, role-play and reenacting help in the acquirement of new skills and the comprehension of complex ideas as per Brinia et al. (2019). Non-negligible, in addition, is the perspective of Moreno, the father of sociodrama, the precursor to psychodrama, who indicates the sociological importance of the group and its internal relations, through the therapeutic concept of "catharsis" as defined by Aristotle, which has to do with conflict resolution in ancient drama (Moreno, 1943). Thus, sociodrama could be used by teachers to resolve personal and cultural conflicts in the classroom, bringing society closer to social inclusion and heightened levels of social and emotional intelligence (Moreno, 1953; Brinia et al., 2018), subsequently increasing social cohesion and therefore social sustainability. It is then obvious that art and culture go hand-in-hand with the course of society (Georgitsogianni, 2011), or as Geertz (2003) would put it, the symbols of culture shape social and moral models. In any case, as the above-mentioned cultural promenades and ancient drama indicate, an aesthetic view on the world leads to improved quality of life and the understanding individuals have on themselves and others, as well as their environment.

Let there, however, be an emphasis on individual identity, at this point. This is a practical matter of the teacher being able to navigate through differences in identity. Teachers play a crucial role in the learning process and, at the same time, enter the classroom with a knowledge background affected by their cultural environment (Salas & López, 2008). Thus, their cultural identity is important, because when they understand their cultural background and connect it to the students, they create a rich learning environment of mutual appreciation (White, Zion, Kozleski & Fulton, 2005). Of course, as teachers themselves have cultural and social identities, those should be respected; as teachers are an integral part of every community, they need to be intellectually involved even for their own professional development (Yogev & Michaeli, 2011; Hughes, 1991). The educators, on the other hand, should keep in mind that each student comes with different styles of cultural learning that should be understood, respected, encouraged, and incorporated in teaching (Hilliard III, 1989). Educators act as mentors who answer students' questions, motivate them, coordinate their cultural experience, encourage them to participate in a team, manage the dynamic relationships formed, and help them develop interpersonal skills (Denham, Ji & Hamre, 2010; Brinia & Athanasiou, 2018; Brinia & Psoni, 2016, 2018, 2019).

The abovementioned fit with goals of Agenda 21 for education and promote sustainability, especially when bringing to mind the importance of teacher leadership, consisting of supporting the classroom to successfully carry out tasks, as well as maintaining effective in-group relationships. This is why Johnson and Johnson (1991) presents keys for group success, both group-maintenance leadership and task leadership, and the requirements for responsible leadership, flexible behavior, the ability to perceive the behaviors more suiting to particular occasions for the group to function efficiently, and the ability to act on those behaviors or get others to act on them.

A practical example of classroom leadership that embraces cultural identity can be found in the workshop reported by Scheie and Brinia (2017), which was the result of a cooperation between the Teacher Education Program of the Athens University of Economics and Business (TEP-AUEB) and the Department of Teacher Education and School Research at the University of Oslo (ILS, UiO). At first, a lecture on leadership was conducted in AUEB, and subsequently, there was an experimental workshop in the New Acropolis Museum, which sought whether classroom leadership principles can be traced back to ancient Greek culture through 5th-century BC statues, aiming at enhancing Greek teacher-candidates' cultural identity in a favorable direction towards leadership. Indeed, Greek culture and leadership meet for the teacher-candidates to improve acquiring self-knowledge and cognitive knowledge, while teamwork, development of interpersonal skills, and critical thought were promoted. Besides, as White et al. (2005) would have it, teachers who understand and value their cultural identity (White, Zion & Kozleski, 2005), and share them together with their lives and life experiences (Eleuterio, 1997; Hoelscher, 1999), build stronger learning environments, relationships, and trust with the students.

Such a conceptual cluster of education, cultural heritage, and sustainability should include social entrepreneurship, but there is limited work on that matter, which is a surprise when considering that entrepreneurship is a crucial driving factor for economic development in macro- and microeconomic terms (Gorman, Hanlon & King, 1997; Bruyat & Julien, 2000; Henry, Hill & Leitch, 2018), while education is the long-term key of economic development and entrepreneurship (Reynolds et al., 2002; Solomon, Duffy & Tarabishy, 2002). At the same time, the core of entrepreneurship lies in opportunity recognition (Eckhardt & Shane, 2003; Austin, Stevenson & Wei-Skillern, 2006; Mair & Marti, 2006), which, however, bears particular characteristics when social entrepreneurship is concerned (Dees, 2007; Dorado, 2006), as it seeks to create value for society (Thompson, 2002), and can bring people closer to their cultural heritage, while it can trigger the socioeconomic regeneration of rural areas (Stobart & Ball, 1998), through economic and social results, such as increased employment, reduced social exclusion, renewed communal feelings and liveliness, independence, and empowerment (Dumas, 2001). There are also environmental outcomes concerning physical infrastructure, which improves the landscapes for locals and tourists (Lindgren & Packendorff, 2003) with a great concern reserved on avoiding negative externalities and combining business viability, well-distributed social inclusion, and sustainability on the ecological level.

Such a need for the entrepreneurial skills in the triangle of education, cultural heritage, and sustainability is a desideratum, which is not often addressed and which is crucial for real-life, applied results, delivering solutions on real-life problems, from a human-centric perspective through the customer-centric, as well as viable, productive structures. Such kinds of actions go through education, more specifically inquiry-based learning and project-based pedagogy that seek to teach through posing real-life problems and calling for a holistic approach by the students, who produce an actual proposal on the problem (Friesen & Scott, 2013). One of the few examples that respond to the problem of sustainability through education and cultural heritage, but via the inquiry-based approach, is the "Education, Entrepreneurship and Cultural Heritage" Initiative, a cooperation of the Teachers Education Program of the Athens University of Economics and Business and the NGO Diazoma (n.d.), which through a cultural promenade and a subsequent project by task groups of teacher-candidates resulted in student proposals of viable development for Orchomenos in Greek Boeotia, complete with business plans and even some innovative suggestions that solve the problem of personalization through digital means, augmented reality, and gamification (Han, Jung & Gibson, 2013; Landers, 2014; Kapp, 2012). The fine work of Diazoma with even more actions and cultural promenades and the prospect of further cooperation between institutions and the use of digitalization and gamification are also described in Brinia and Belogianni (2019).

3.5 CONCLUDING REMARKS

All the above, then, showcase how education and cultural heritage can work together for sustainability and how they already include it in their "character." It is, through experiential and project-based learning, social entrepreneurship, an understanding of the importance of cultural identity, and interdisciplinary approaches and initiatives, that this schema comes together. The aim of this innovative perspective is that through this three-part synergy, the improvement of people's quality of life and their empowerment are achieved, and economic development is enhanced in the long run, with minimal negative externalities and with high social cohesion and social inclusion.

REFERENCES

Antonacopoulou, E. P. & Gabriel, Y. (2001). Emotion, learning and organizational change: Towards an integration of psychoanalytic and other perspectives. *Journal of Organizational Change Management, 14*(5), 435–451.

Aristotle. (2011). Nicomachean Ethics, *Book Two*. Translated Chase, D. P. Retrieved from: https://en.wikisource.org/wiki/Nicomachean_Ethics (Chase)/Book_Two.

Austin, J., Stevenson, H., & Wei-Skillern, J. (2006). Social and commercial entrepreneurship: Same, different, or both? *Entrepreneurship Theory and Practice, 30*(1), 1–22.

Bassi, L. J. & Russ-Eft, D. F. (1997). *Assessment, Development, and Measurement* (Vol. 1). Alexandria, VA: American Society for Training and Development.

Bloom, B. S. (1956). *Taxonomy of Educational Objectives, Handbook* Cognitive domain (Vol. 1, pp. 20–24). New York: McKay.

Borsay, P. (1986). The rise of the promenade: The social and cultural use of space in the English provincial town c. 1660–1800. *Journal for Eighteenth-Century Studies, 9*(2), 125–140.

Bourdieu, P. (1986). Forms of capital. In: J. G. Richardson (ed.) *Handbook of Theory and Research for the Sociology of Education* (pp. 241–258). New York: Greenwood Press.

Brinia, V. & Athanasiou, A. (2018). Teachers' views and attitudes on the organization and implementation of the social and emotional intelligence education program. *Education Quartely Review, 1*, 47–65.

Brinia, V. & Belogianni, M. (2019). Epicheirimatikotita, Ekpaideusi kai Politismos. Synergasia OPA me DIAZOMA [Entrepreneurship, Education and Culture. Cooperation of AUEB with DIAZOMA]. *Xenophon, 4*, 121–129.

Brinia, V. & Psoni, P. (2016). Educators' perceptions about incentives and their role in students' learning. *International Journal of Academic Research in Progressive Education and Development, 5*(4). Retrieved from: http://hrmars.com/hrmars_papers/Educators%E2%80%99_Perceptions_about_Incentives_and_their_Role_in_Students%E2%80%99_Learning.pdf.

Brinia, V. & Psoni, P. (2018). Multi-level mentoring practices in a Teacher Education Program in Greece: How their effectiveness is perceived by mentors. *Journal of Applied Research in Higher Education, 10*(3), 256–270.

Brinia, V. & Psoni, P. (2019). How to develop cognitive skills through playing in pre-school contexts. *International Journal of Teaching and Case Studies, 10*(1), 1–11.

Brinia, V., Psoni, P., & Ntantasiou, E. K. (2019). How to instill cultural values in the new generation through cultural promenades and ancient drama: A field research. *Sustainability, 11*(6), 1758.

Bruyat, C. & Julien, P. A. (2000). Defining the field of research in entrepreneurship. *Journal of Business Venturing, 16*(2), 165–180.

Costanza, R. & Daly, H. E. (1992). Natural capital and sustainable development. *Conservation Biology, 6*(1), 37–46.

Cannon, H. M. & Feinstein, A. H. (2014). Bloom beyond Bloom: Using the revised taxonomy to develop experiential learning strategies. *In Developments in Business Simulation and Experiential Learning: Proceedings of the Annual ABSEL Conference* (Vol. 32), Orlando, FL.

Chiu, R. L. (2004). Socio-cultural sustainability of housing: A conceptual exploration. *Housing, Theory and Society, 21*(2), 65–76.

Dees, J. G. (2007). Taking social entrepreneurship seriously. *Society, 44*(3), 24–31.

Denham, S. A., Ji, P., & Hamre, B. (2010). Compendium of preschool through elementary school social-emotional learning and associated assessment measures. Collaborative for Academic, Social, and Emotional Learning.

Dewey, J. (2005). *Art as Experience*. London: Penguin Books.

Dorado, S. (2006). Social entrepreneurial ventures: Different values so different processes of creation, no? *Journal of Developmental Entrepreneurship, 11*(4), 319–343.

Dumas, C. (2001). Evaluating the outcomes of micro-enterprise training for low income women: A case study. *Journal of Developmental Entrepreneurship, 6*(2), 97–129.

Eckhardt, J. & Shane, S. (2003). Opportunities and entrepreneurship. *Journal of Management, 29*(3), 333–349.

Eleuterio, S. (1997). Folk culture inspires writing across the curriculum. *C.A.R.T.S. Newsletter, 4*.

European Commission. (2005). Impact Assessment Guidelines SEC 791. Retrieved from: https://ec.europa.eu/transparency/regdoc/rep/2/2005/EN/SEC-2005-791-2-EN-MAIN-PART-1.PDF.

Friesen, S. & Scott, D. (2013). Inquiry-based learning: A review of the research literature. *Alberta Ministry of Education, 69*, 1–32.

Geertz, C. (2003). Thick description: Toward an interpretive theory of culture. *Culture: Critical Concepts in Sociology, 1*, 173–196.

Georgitsogianni, E. (2011). *Eisagogi Stin Istoria tou Politismou [Introduction to the History of Culture]*. Athens: Diadrasi.

Gibson, R. B. (2001). *Specification of Sustainability-Based Environmental Assessment Decision Criteria and Implications for Determining "Significance" in Environmental Assessment*. Ottawa: Canadian Environmental Assessment Agency.

Gorman, G., Hanlon, D., & King, W. (1997). Some research perspectives on entrepreneurship education, enterprise education and education for small business management: A ten-year literature review. *International Small Business Journal, 15*(3), 56–77.

Han, D. I., Jung, T., & Gibson, A. (2013). Dublin AR: Implementing augmented reality in tourism. In: Z. Xiang & I. Tussyadiah (Eds.) *Information and Communication Technologies in Tourism* 2014 (pp. 511–523). Cham: Springer.

Harfitt, G. J. & Chow, J. M. L. (2018). Transforming traditional models of initial teacher education through a mandatory experiential learning programme. *Teaching and Teacher Education, 73*, 120–129.

Helming, K., Pérez-Soba, M., & Tabbush, P. (Eds.) (2008). *Sustainability Impact Assessment of Land Use Changes*. Berlin: Springer Science & Business Media.

Henry, C., Hill, F., & Leitch, C. M. (2018). *Entrepreneurship Education and Training*. London: Routledge.

Hilliard III, A. G. (1989). Teachers and cultural styles in a pluralistic society. *NEA Today*, *7*(6), 65–69.

Hoelscher, K. J. (1999). Cultural watersheds: Diagramming one's own experience of culture. *Social Studies and the Young Learner*, *12*(2), 12–14.

Hotelling, H. (1931). The economics of exhaustible resources. *Journal of Political Economy*, *39*(2), 137–175.

Hu, H., Real, E., Takamiya, K., Kang, M. G., Ledoux, J., Huganir, R. L., & Malinow, R. (2007). Emotion enhances learning via norepinephrine regulation of AMPA-receptor trafficking. *Cell*, *131*(1), 160–173.

Hughes, P. (1991). Teachers' Professional Development. Teachers in Society Series. Customer Services, Australian Council for Educational Research, PO Box 210, Hawthorn, Victoria, 3122.

Isyar, Ö. Ö. & Akay, C. (2017). The use of "drama in education" in primary schools from the viewpoint of the classroom teachers: A mixed method research. *Online Submission*, *8*(28), 201–216.

Hawkes, J. (2001). *The Fourth Pillar of Sustainability: Culture's Essential Role in Public Planning*. Melbourne: Common Ground.

Johnson, D. W. & Johnson, R. T. (1991). *Learning Together and Alone*. Englewood Cliffs, NJ: Prentice-Hall.

Kapp, K. M. (2012). Games, gamification, and the quest for learner engagement. *T + D*, *66*(6), 64–68.

Kolb, D. A. (2014). *Experiential Learning: Experience as the Source of Learning and Development*. Englewood Cliffs, NJ: FT Press.

Kuhlman, T. & Farrington, J. (2010). What is sustainability? *Sustainability*, *2*(11), 3436–3448.

Landers, R. N. (2014). Developing a theory of gamified learning: Linking serious games and gamification of learning. *Simulation & gaming*, *45*(6), 752–768.

Lindgren, M. & Packendorff, J. (2003). A project-based view of entrepreneurship: Towards action-orientation, seriality and collectivity. In: C. Steyaert & D. Hjorth (Eds.) *New Movements in Entrepreneurship* (pp. 86–102). Cheltenham: Edward Elgar.

Mair, J. & Marti, I. (2006). Social entrepreneurship research: A source of explanation, prediction, and delight. *Journal of World Business*, *41*, 36–44.

Malthus, T. R. (1986). An essay on the principle of population. 1798. *The Works of Thomas Robert Malthus, London, Pickering and Chatto Publishers*, *1*, 1–139.

McLeod, S. A. (2017). Kolb-learning styles. Retrieved from: www.simplypsychology.org/learning-kolb.html.

Moss, G. (2008). Diversity study circles in teacher education practice: An experiential learning project. *Teaching and Teacher Education*, *24*(1), 216–224.

Moreno, J. L. (1943). The concept of sociodrama: A new approach to the problem of inter-cultural relations. *Sociometry*, *6*(4), 434–449.

Moreno, J. L. (1953). *Who Shall Survive? Foundations of Sociometry, Group Psychotherapy and Socio-Drama* (2nd edn). Oxford: Beacon House.

Nassauer, J. I. (1997). *Cultural Sustainability: Aligning Aesthetics and Ecology*. Washington, DC: Island Press.

Naveh, Z. (2007). Landscape ecology and sustainability. *Landscape Ecology*, *22*(10), 1437–1440. doi: 10.1007/s10980-007-9171-x.

Nurse, K. (2006). Culture as the fourth pillar of sustainable development. *Small States: Economic Review and Basic Statistics*, *11*, 28–40.

O'Byrne, A. F. (2006). Walking, rambling, and promenading in eighteenth-century London: A literary and cultural history (England).

Pine, B. J. & Gilmore, J. H. (1998). Welcome to the experience economy. *Harvard Business Review, 76*, 97–105.

Poundstone, W. (1992). *Prisoner's Dilemma.* New York: Doubleday.

Priest, S. (1986). Redefining outdoor education: A matter of many relationships. *The Journal of Environmental Education, 17*(3), 13–15.

Randall, A. (2008). Reflections on Solow's Richard T. Ely Address. *Journal of Natural Resources Policy Research, 1*(1), 97–101.

Reynolds, P. D., Bygrave, W. D., Autio, E., Cox, L. W., & Hay, M. (2002). Global entrepreneurship monitor: Executive report, GEM, Babson College and Ewing Marion Kaufmann Foundation, Wellesley, MA and Kansas City, MO.

Robert, K. W., Parris, T. M., & Leiserowitz, A. A. (2005). What is sustainable development? Goals, indicators, values, and practice. *Environment: Science and Policy for Sustainable Development, 47*(3), 8–21.

Robbins, D. (1991). *The Work of Pierre Bourdieu: Recognizing Society.* Boulder, CO: Westview Pr.

Salas, L. & López, E. J. (2008). Cultural identity and special education teachers: Have we slept away our ethical responsibilities? *Journal of the American Academy of Special Education Professionals, 47*, 53.

Scheie, J. & Brinia, V. (2017). Cultural identity as a tool for classroom management. *International Journal of Multidisciplinary Education and Research, 2*(4), 67–71. Retrieved from: www.educationjournal.in/download/227/2-4-47-507.pdf.

Solomon, G. T., Duffy, S., & Tarabishy, A. (2002). The state of entrepreneurship education in the United States: A nationwide survey and analysis. *International Journal of Entrepreneurship Education, 1*(1), 65–86.

Stavridis, S. (2006). *Mnimi Kai Empeiria tou Chorou [Memories and Experiences of the Landscape].* Athens: Alexandreia.

Stobart, J. & Ball, R. (1998). Tourism and local economic development. *Local Economy, 13*(3), 228–238.

Strange, T. & Bayley, A. (2008). *Sustainable Development-Linking Economy, Society, Environment.* Paris: Organization for Economic Co-Operation and Development.

Terkenli, T. (1996). *To Politismiko Topio: Geografikes Proseggiseis [The Cultural Landscape-Geographical Approaches].* Athens: Papazissis and University of the Aegean.

Tibbs, H. (2011). Changing cultural values and the transition to sustainability. *Journal of Futures Studies, 15*(3), 13–32.

The "Education, entrepreneurship and cultural heritage" Initiative | Athens University of Economics and Business (n.d.). Retrieved from: https://www.dept.aueb.gr/en/tep/content/%E2%80%9Ceducation-entrepreneurship-and-cultural-heritage%E2%80%9D-initiative.

Thompson, J. (2002). The world of the social entrepreneur. *International Journal of Public Sector Management, 15*(5), 412–431.

Throsby, D. (1999). Cultural capital. *Journal of Cultural Economics, 23*(1–2), 3–12.

Throsby, D. (2001). *Economics and Culture.* Cambridge: Cambridge University Press.

Throsby, D. (2008). Linking cultural and ecological sustainability. *International Journal of Diversity in Organisations, Communities and Nations, 8*(1), 15–20.

UNESCO. (2002). Education for sustainability from Rio to Johannesburg: Lessons learnt from a decade of commitment. Retrieved from: https://unesdoc.unesco.org/ark:/48223/pf0000127100.

United Nations Agenda for Development. (1997). Department of Public Information. Retrieved from: www.un.org/documents/ga/res/51/ares51-240.htm.

WCED, S. W. S. (1987). World commission on environment and development. *Our Common Future, 17*, 1–91.

Weber, M. (2013). *The Protestant Ethic and the Spirit of Capitalism.* New York: Routledge.

Weiss, R. P. (2000). Emotion and learning. *Training and Development, 54*(11), 45.

Wiersum, K. F. (1995). 200 years of sustainability in forestry: Lessons from history. *Environmental Management, 19*(3), 321–329.

Wilderer, P. A. (2007). Sustainable water resource management: The science behind the scene. *Sustainability Science, 2*(1), 1–4.

White, K., Zion, Sh., & Kozleski, E. (2005). Cultural identity and teaching. National Institute for Urban School Improvement. Arizona State University.

White, K. K., Zion, S., Kozleski, E., & Fulton, M. L. (2005). *Cultural Identity and Teaching.* Tempe, AZ: National Institute for Urban School Improvement.

Yogev, E. & Michaeli, N. (2011). Teachers as society-involved "organic intellectuals": Training teachers in a political context. *Journal of Teacher Education, 62*(3), 312–324.

4 "I Can Be the Beginning of What I Want to See in the World"

Outcomes of a Drama Workshop on Sustainability in Teacher Education

Eva Österlind
University of Stockholm

CONTENTS

4.1   INTRODUCTION

There is a momentum for a huge, joint effort on a global scale to (re)create sustainable ways of living on this planet. This giant challenge needs to be acknowledged at all levels of society, from world leaders' strategic decision-making to individual consumers' choices. All forms of education, from preschools to universities, have been identified as having a key role. This was explicitly stated when the UN declared a Decade of Education for Sustainable Development (2005–2014). The purpose was to mobilize the educational resources of the world in order to contribute to a more sustainable future (UNESCO, 2005). More recently, the UN have launched Sustainable Development Goals (SDGs), covering and connecting many sectors in our societies (UNESCO, 2017). These goals are presented in order to guide and strengthen strategic planning, decisions, and actions for sustainability worldwide, also in education.

Overarching and generally accepted but slightly vague concepts like "democracy" or "sustainability" are often supposed to permeate all education – usually without any specific goals, priorities, means, or resources. As a consequence, to embrace and put forward these ideas and concepts may end up as individual, committed teachers' responsibility, or in worst case nobody cares and teaching concerning these core concepts tends to evaporate. Introducing a new subject area is seldom a simple, straightforward process, at any educational level. This is obvious when it comes to SDS in higher education (HE), where the inclusion of SDGs in universities is in its "infancy" and more systematic efforts are needed (Leal Filho et al., 2019).

There are several strategies for implementing SDGs in HE. The most evident is to design a specialized course or a degree in or closely related to sustainable development. This is crucial for knowledge accumulation and qualification and should by no means be diminished. Still, the vast majority of all students are enrolled in other areas (e.g., economy, nursing). In such cases, the best option might be to "feed in" some aspects of sustainability, by connecting it to other subjects and themes, whenever there is a possibility. This is much better than nothing, although inherent problems are, for instance, lack of continuity and progression. Other challenges connected to HE have also been identified (UNESCO, 2016).

Consequently, there is a body of research concerning teaching and learning for sustainability (environmental education, education for sustainable development/ sustainability) within HE (e.g., Díaz-Iso et al., 2019; Greig & Priddle, 2019; Zamora-Polo & Sánches-Martín, 2019). One example is Álvarez-García et al. (2019), who studied several factors that might affect pre-service student-teachers' achievement of environmental competencies, including knowledge, attitude, and behavior. The results showed, among others, that former studies, average grade level, or parents' education did not have any impact. This result is very promising, as those factors are beyond reach for university teachers. What turned out to have a considerable impact was if the students had taken university courses connected to sustainability (e.g., science, health, and sustainability). This is not too surprising, but as I read the result, it underlines the importance of including sustainability in HE, whenever possible and relevant, even if the circumstances are not ideal.

As education is considered to have a key role in sustainability, teacher education becomes particularly relevant. In the context of teacher education, there is usually at least a double purpose. In all education, students are supposed to increase their knowledge about a particular subject area and, by doing so, widen their perspective and gain something for themselves (cf. bildung). In teacher education, the purpose is also to learn about teaching. This includes teaching in general (e.g., classroom management, establishing an open atmosphere) and how to teach a specific subject. These purposes are hard to separate, but either personal or professional development, and general or subject-specific knowledge, can be in the foreground on different occasions.

Just like a teacher, education has a double agenda: "I have an interest in educational drama for active, inclusive, embodied, and interactive learning, and an interest in teaching for sustainable development." This paper is dedicated first and foremost to contribute to new, more relevant ways of teaching for sustainability. In other words, this is a research-based project for educational development. As a consequence, the historical variations and ongoing debates about how to label the hot topic of sustainability, and the finer nuances in how to entitle drama work, are left aside. Likewise, theories of learning (e.g., Mezirow, 2000) are not in focus here. So, how can applied drama contribute to learning for sustainability in HE? Given that overarching question, it is evident that the two aspects can't be completely separated, but here, learning for sustainability is at the center, and drama work is seen as the multimodal medium.

4.2   PREVIOUS RESEARCH AND PURPOSE

Sustainability challenges can be understood as a result of alienation, separating social and economic systems from nature, and separating decision-making from emotions and values (Laininen, 2018). To counteract this, we need to "re-connect," by fostering a deeper, relational understanding of the interdependence between nature and society, the connection between local and global levels, and so forth. Consequently, there is a request for new ways of teaching and learning in education for sustainability (ESD) in HE (Lotz-Sisitka et al., 2015). For instance, more holistic teaching approaches are needed (Cantell et al., 2019), emphasizing our interconnectedness (Lehtonen et al., 2018).

There is no agreement regarding how to define sustainability, and perhaps it is not necessary or even possible to find a generally accepted definition. Instead, the meaning of sustainability can be seen as constantly processed and negotiated (Christie et al., 2013). As a concept, sustainability is characterized by "ambiguity, uncertainty, and open-endedness" (Eernstman & Wals, 2013, p. 1647). These fairly unstable characteristics of sustainability, combined with its obvious, transdisciplinary features, make teaching for sustainability significantly different from teaching more traditional academic content in HE.

Some researchers argue that in ESD "we have to attend to the value judgements before we can attend to the facts" (Lundegård & Wickman, 2007, p. 14), and that "we have to facilitate coping with the calamities around us – both those that are manifest and those that are feared for" (van Boeckel, 2009, p. 145). As if that was not enough, the inherent dilemma between imposed changes in lifestyle or lack of sustainability makes the topic value loaded and emotionally challenging, for teachers as well as students. Young people are often ambivalent or uncertain in relation to environmental problems (Ojala, 2012a), and both children and adults are worried or even frightened (Österlind, 2012, 2018). This may lead to psychological defenses, like reluctance, but to avoid hopelessness and passivity, it is important to focus on possible solutions (Ojala, 2012b).

Taken together, sustainability is a controversial and challenging subject area in HE. As a consequence, teachers may look for "safe ground" that everyone can agree on (e.g., recycling). This is understandable but leads nowhere. On the contrary, Læssøe recommends to work with dilemmas and dissensus in ESD, and points out the risk that ESD oriented towards consensus and minor changes leads to "societal self-deception" (Læssøe, 2010, p. 51). According to UNESCO (2017), ESD should encourage participants to not only reflect on their own behavior, but to take action and move their societies in a more sustainable direction. This is reasonable, given the global threats, but nevertheless quite hard to achieve and the expectations put some pressure on teachers at all levels.

Given the challenges in teaching for sustainability, there is a request for new educational practices as "practitioners are in the dark when it comes to developing strategies to implement ESD" (Eliason Bjurström, 2012, p. 19). There are many attempts to address these challenges, for example, science education for activism (Roth, 2014), outdoor education like excursions (Tooth & Renshaw, 2009), and work-based learning for professional development (Wall & Hindley, 2018). Within teacher education, suggestions include inquiry-based learning and gamification (Zamora-Polo & Sánches-Martín, 2019), and extra-curricular activities (Días-Iso et al., 2019), just to mention a few.

Arts-based teaching is increasingly recognized as significant in ESD, not least in HE (van Boeckel, 2013; Wall et al., 2018a). For example, UNESCO advocates the application of "arts education principles and practices to contribute to resolving the social and cultural challenges facing today's world" (2010, p. 8). One reason for this recommendation may be that arts-informed teaching is based not only on logic and rationality, but also on knowing derived from the body and senses. Within an "aesthetic paradigm," cognitive, affective, and embodied aspects are not seen as opposed or contradictive, but as mutually dependent and complementary forms of knowledge and learning, or in other words a fusion of thinking, feeling, and doing (cf. Dewey, 1934).

Hunter et al. (2018) put forward arts education as a possibility for students to explore change, while developing their own capacity to actually make change (cf. Österlind, 2008), and suggest to focus on *how* students learn, rather than on *what* they learn about sustainability (see also Glasser, 2018). Applied drama can be one possible response to the call for other forms of teaching, as it offers opportunities to link diverse sources of scientific knowledge with personal experiences, emotion, and ethical judgments (Heras & Tàbara, 2014, p. 379). According to Forgazs (2013), drama has a potential to foster empathy and connectedness to other persons, but also to other species and the planet, as a driving force for action. Drama work is embodied, interactive, and reflective. It addresses both affective and cognitive aspects of learning, and it is action oriented. Thus, a drama workshop differs a lot from a university lecture.

4.2.1 RESEARCH ON DRAMA AND ROLE-PLAY IN HIGHER EDUCATION FOR SUSTAINABILITY

Drama (applied drama, educational drama, drama in education) is an arts-based teaching approach based on play, stories, and theater. It is closely related to applied theater, although in drama work there is usually no manuscript or rehearsal, and no external audience. Instead, drama is based on improvised interaction in a fictional context, where sometimes devised scenes are played in front of the other participants. Educational drama work makes use of traditional games and exercises developed for actor training, always adapted to the particular context. The components, or "drama conventions," can be combined in many ways, depending on the purpose, the circumstances, and the participants at any given occasion. Drama work includes embodied and verbal interaction and reflection, and is often designed to experience different perspectives and dilemmas (Wall et al., 2018b). It allows participants to explore real problems, but from a safe position in a fictive situation (Heyward, 2010). Educational drama is process oriented and can be described as a chain of carefully planned tasks and unfolding, unpredictable events (Österlind, 2018).

Role-play is an integrated part of the larger drama-teaching repertoire, but is also applied separately in a wide range of academic disciplines. A role-play is usually designed to make sure that different stakeholders and perspectives meet in a "structured controversy" to learn from (Wareham et al., 2006, p. 651). The format of a role-play varies considerably, depending on the context and the teacher's knowledge and experience. For example, a role-play in HE can be the outcome of a long-term commitment or a single occasion. It can be based on careful preparation or an ad hoc activity. It may go on for 2 days, or half an hour. The structure can be relatively strict, even based on a script, or quite open, and flexible. It can focus on academic content knowledge or embodied self-reflection. When a role-play takes place in a drama context, the design is more likely to focus on personal reflections, empathy, and embodied experiences, generated by "stepping into someone else's shoes."

There are several studies of role-play, but only a few studies of drama in ESD in HE. Below, studies of drama and/or role-play in ESD/environmental studies in HE are presented, without distinction. Such studies are connected to disciplines like biology (Oliver, 2016), engineering (Wareham et al., 2006), environmental education (Chen & Martin, 2015), and geography (Schnurr et al., 2014), but also to teacher education (Álvarez-García et al., 2019) and work-based learning and professional development (Österlind, 2018).

In general, there are two models for implementing drama/role-play in ESD in HE. One model is less time-consuming, and more of an ad hoc event – which of course doesn't mean it isn't carefully planned. The students' time for preparation may be only 20 min. (Wareham et al., 2006), and the role-play is a single event (Blanchard & Buchs, 2015). It can also take place in a drama context, with almost no time for students to prepare (Österlind, 2018).[1] This model is often applied as an

---

[1] Very little time for preparation is not only due to time limits, but can be a strategy to lower the pressure on students who might think that they have to "act theatrically," which is not the case (cf. Gordon & Thomas, 2016).

introduction to ESD or to a specific concept, either in the beginning of a course or as an occasional event.

The other model is more time-consuming and may be more demanding. The students' preparation (e.g., gather information, create a strategy) may go on for several weeks (Paschall & Wüstenhagen, 2012), or at least a few days, and may include written assignments (Buchs & Blanchard, 2011). This model usually appears towards the end of a course, when the students have gained a certain level of knowledge that they now get an opportunity to apply (and demonstrate). In both models, the role-play ends with a reflective evaluation or "debriefing session" (Blanchard & Buchs, 2015). This final, reflective part is absolutely crucial from a learning perspective, no matter if the role-play is the outcome of careful preparation or more based on improvisation.

The reported results are often positive. Buchs and Blanchard (2011) developed a role-play on the concept of sustainability, which was conducted several times and reached many students. Nearly 80% of 179 students reported that the role-play clearly supported their critical thinking, and more than 90% said that it led to changed behavior. Buchs and Blanchard conclude that the role-play enabled the students to "understand the concept of sustainability in a less normative format" (2011, p. 709). Gordon and Thomas (2016) found that even though role-play is time-consuming and often demanding, it is definitely worthwhile. One of their students once said: "the learning sticks" (p. 14).

In a study about "social sustainability" in the engineering curriculum, Björnberg et al. (2015) concluded that study visits and role-play were most effective. Cruickshank and Fenner (2012) compared several "student-centered activities" in a master's program on engineering for sustainable development. They also found that field courses and role-play exercises had the strongest impact.[2] In a study of how to reinforce connections between university students and the environment, Davis and Tarrant (2014) integrated science-based, fictive and experiential components, including drama and applied theater. They found that a combination of outdoor experiences and drama work was especially effective.

Even though the results are mainly positive, some difficulties are also put forward. Role-playing can be time-consuming (Gordon & Thomas, 2016), and sometimes it takes a lot of staff resources (Paschall & Wüstenhagen, 2012). Gordon and Thomas (2016) also mention that students may feel anxious before they realize that role-playing is not about formal acting. Davis and Tarrant (2014) point out a tension between the need for facts and research-based teaching for sustainability, and drama work which is affective, embodied, and takes place in the fictive realm.

4.2.2 PURPOSE

The urgency of global warming and other environmental challenges has led to a call for new, more relevant ways of teaching for sustainability. There is a need to know more about how to teach ESD, not least in HE, as today's students most likely

---

[2] In a study of children and youth, Ballantyne and Packer (2007) compared "experience-based strategies" for environmental education. Even in this case, "outdoor education" and "story or drama" got the highest scores.

will be involved in the necessary transformation. As the majority of the students are studying in other academic areas, there may be very limited time allocated to learn for sustainability. Teachers and teacher education are considered to have a key role in ESD. Leaving campus for outdoor experiences or field trips seems to have a considerable impact but can be difficult to arrange. Research clearly indicates that drama and role-play may contribute significantly, although the amount of studies is still limited. So, how can drama contribute to ESD in HE? What can be achieved with a single drama workshop on sustainability in teacher education? These overarching questions will be discussed based on the outcomes of the following research questions:

1. How do student-teachers experience and reflect on educational drama work?
2. How do student-teachers reflect on sustainability after one drama workshop?

4.3 METHODOLOGY

The empirical material is based on a questionnaire concerning university students' experiences of a drama workshop on sustainability, given to student-teachers in Athens. The same workshop and questionnaire have been given to postgraduate students in Helsinki (Österlind, 2018). The previous study will serve as a point of reference to widen the perspective, and to some extent add to the discussion of the present study. In Athens, ~60 students were invited to participate in a drama workshop as part of their teacher education. These students already had a bachelor's degree in economy and were ready to graduate as upper secondary school teachers. The workshop was given by the author, as part of an academic exchange program between Athens University of Economics and Business, and Stockholm University.

The intervention had more than one purpose. One aim was to exemplify arts-based teaching in HE, by introducing drama work in a formal educational context. The workshop could be seen as a demonstration of educational drama work, but the main intention was to offer the participants a first-hand experience, to find out for themselves what applied drama work can be like. A second purpose was obviously to put forward issues of sustainability. A third purpose was to suggest a certain way to address ESD, which from a teaching perspective can be quite complicated and demanding.

Directly after the workshop, the students were asked to fill in a questionnaire, including three open-ended questions. The students' responses to these open questions constitute the data material presented here. It is a weakness that the students were asked to fill in the questionnaires immediately after the workshop, for two reasons. One is that the experience was very "fresh" and there was no break for individual reflection. The other is that we don't know anything about the impact after a period of time, for instance, a month or even a year (cf. Österlind, 2018). Conducting the study in this way was related to logistic and organizational conditions. Another possible limitation, depending on the research interest, is that the study is solely based on students' descriptions of their own experiences. If the student perspective matches the scope of the study, as in this case, it is not a problem. The three open questions were as follows:

a. What do you think about the workshop we just did?
b. Did you learn anything from the workshop? If so, please tell us what you learned! (About environmental problems/sustainability? About drama? About yourself/others?)
c. What do you think about using drama or doing something like this in your own teaching? (Possibilities? Difficulties? Students' responses?).

Of the 58 students, two students did not respond to the open questions. It is worth mentioning that the students were asked to answer in English, which for the Greek students means not only writing in a foreign language, but also using a less familiar alphabet. Despite this, 56 students responded to the three open questions, and many answers were quite rich and elaborated. The students' responses were not interpreted, looking for underlying perspectives or implicit messages. Instead, attention was paid to their actual, written responses.

The qualitative data material was analyzed by using content analysis, searching for themes emerging in the data (Elo et al., 2014), and applying a standard coding process (Bogdan & Biklen, 2003). This is done in order to extract themes from the data, which can be described as meaning coding and meaning condensation (Kvale, 2007). The inductive process started with a very open approach, to find out what the students had written about. The following steps were also fairly open, not tied to the questions in the questionnaire, but slowly more in line with the two research questions presented earlier.

The students' responses were examined by looking for certain areas of interest: How did the students experience the drama workshop? Do they reflect on sustainability issues? What are their views on using drama in their own future teaching, in general and/or for sustainability? These questions can be structured according to two related aspects. One is the dimension between personal experiences and learning related to the future profession. The other aspect is connected to the content, if students reflect on the drama work as such, on the sustainability theme, or both. The possible connections between these aspects are presented in Table 4.1.

Because of the research interest, but mainly due to the students' responses, the analysis concerns only two of these areas. Students' personal perspective on learning

TABLE 4.1

Possible and Actual Combinations of Areas of Interest in Students' Responses

Focus	Educational Drama Work	Sustainability Issues	Drama Work and Sustainability
Personal perspective Reflections on learning experience			X
Professional perspective Reflections on future teaching	X		

experiences related to drama and sustainability, and their professional perspective related to using drama in their own future teaching are central. This means that students' personal reflections on drama work in general, and their professional reflections on sustainability teaching, in general, are regarded as more peripheral in this context, which also do justice to the data material. As Table 4.1 indicates, all students reflected on the drama work, and there were no students who only reflected on sustainability issues.

Before turning to the results, the drama workshop will be briefly described. The workshop, which lasted approximately 3 h, was originally designed as an example of drama teaching for student-teachers with no previous drama experience. The first part of the workshop concerns individual and collective reflection on sustainability challenges. It includes brainstorming some of the main problems for the globe, followed by guided relaxation and individual introspection in silence; "what do you feel in front of these problems?" It also includes sharing ideas and thoughts in pairs or in small groups, making (bodily) still images about the human causes of the global problems, and joint reflections of what needs to be done and by whom. The second part is based on a role-play, a fictive environmental conference, designed to look at climate change from different stakeholders' perspectives. The students join the conference in teams (e.g., activists, business leaders, politicians, researchers). The teacher is in role as the hostess. At the conference, each team presents their interests related to environment and climate change and discusses the issues in mixed groups. After some reflections, in role and out of role, the workshop ends by guided relaxation and introspection. Everyone is asked to think of something they are able to do for these global problems, and to privately make a decision if they are ready to actually do it. One purpose is to face the global problems and connect them to the personal level. Another purpose is to address the question of responsibility, at all levels. The overarching purpose is increased awareness, and in the best case a push for taking action.

4.4   RESULTS AND DISCUSSION

The results are based on questionnaires given to student-teachers in Athens. One question with five given responses concerned if the student was familiar or unfamiliar with drama. Two-thirds, or 38, of the students were unfamiliar with drama, ten were familiar with drama, and ten considered themselves to be in between. The questionnaire included three open questions, and 56 of the 58 students responded. Of totally 168 possible answers, 160 were given. In other words, the response rate was notably high. The students' responses were read carefully, closely, and repeatedly, while looking for patterns, themes, and dimensions in order to describe and characterize the data material in a representative way. The outcome is presented below, where more detailed descriptions are followed by an overview.

The most obvious dividing line in the material is if the students mention the content or theme of the workshop or not. Out of the 56 students, 15 students do not refer to sustainability at all. This does not imply the content meant nothing to them, but in any case, they did not explicitly write about it. For someone who is unfamiliar with drama, the experience of the multimodal drama work tends to be in the foreground and the theme or content can get slightly out of sight. The other 41 students all refer to the thematic content, usually in terms of "environment." Another obvious dimension, regardless of mentioning the environment or not, concerns the students' different approaches to use drama in their own future teaching. Nearly all students comment positively on the drama experience, but their approaches to apply drama in their own teaching differ significantly, as presented below.

4.4.1 Students' Approaches to Apply Educational Drama

To quickly capture the students' different approaches to educational drama, without making any definite claims, they could be characterized as the *optimistic, realistic,* and *pessimistic* approach. It must be noted that nearly all students are very positive to drama work as such, but differ in their view of using drama in their own teaching. The following description of the students' approaches to apply drama in their own future teaching is based on quotes from the students. The quotes have in some cases undergone minor editing, to facilitate reading.

Nearly half of the respondents (27 students) are very positive or even enthusiastic towards drama work, and when they filled in their questionnaires, they did not mention any problems or difficulties about applying drama in their own teaching: "I think it will be successful," and "I want to adopt this learning by doing." Drama is seen as important, because it makes teaching more interesting for the students: "It's creative and innovative. Students would be really motivated." Drama allows the students "to learn and to have some fun in the same time." One will "try a lot to introduce the students in drama education," while another will "definitely apply [this new way of teaching] in my courses." Someone learned "a lot of useful techniques that I'm going to use as a teacher." Three of these students refer to already have tried some drama work, one "with good effect." Some responses refer to both personal and pedagogical outcomes. Drama is, for instance, described as a chance to interact and to "develop our creativity," and as an interesting method because "you become more sensitive."

Another group consisting of 17 students is equally positive towards drama, but all of them make some pedagogical reflections about possible difficulties to be handled in connection to drama work. Drama takes time, and the teacher has to be well prepared. The activities should have a clear goal and be relevant for the students' level. Even the students need to be prepared. It can be difficult to manage a large group: "Students talk to each other and I can't control the situation." There can also be "organizational difficulties." And someone identifies a need to learn more: "I think I need to study a lot so I can be the mentor in this process." It is worth noting that these anticipated challenges are not taken as a reason to never use drama, but seen as something to take care of. "I think it will be difficult at first, but we can give it a try." "The difficulties in a classroom full of noisy children will be plenty. But I believe, in some occasions it is totally worth it." "It would be very challenging but yes I would love to try this on my own." The reasons for doing these efforts are several; for example, it will "increase students' engagement in schoolwork" and "it is necessary for a teacher to make students learn themselves."

The third group consisting of 12 students is also positive towards drama, but anticipates severe problems. They describe the workshop as interesting, even "unique" and a "nice experience!!!" But the foreseen difficulties of applying drama are many, mainly due to lack of time, and the need for teachers to be very well prepared. Other concerns are related to their future students; they can be unwilling due to stress, or find drama more like "kids games." One also points out that drama needs "a good mood among all students." Some of the hesitation is connected to personal aspects: "I wouldn't feel comfortable to do this in my own teaching," and the lack of adequate training, "Drama will help in teaching, but we [are not] educated. So we don't know how to use drama in teaching." There are also obstacles of another kind. Drama can be very "fun and educative but under the circumstances of the analytical [study] program and strict headmaster, it would be difficult." Someone also mentions a lack of resources and possibilities.

4.4.2 Comments on Students' Approaches to Applying Drama

In general, nearly all students express a positive or enthusiastic view on drama work. The differences are to be found in how they consider the option to use drama in their own future teaching. The optimistic approach, and the enthusiasm for applying drama, could be linked to the immediate experience of taking part in something that was unfamiliar for most of the participants, but turned out to be a quite lively and rewarding activity. Is this enthusiasm, and absence of problems, a sign of being inspired at the moment? Or is it a sign of confidence in one's own capacity as a teacher? Is it connected to already having some drama experience? This is not possible to answer here, but the students who were most familiar with drama share the optimistic approach, while students without any previous drama experience did not adopt this approach. The fact that the students don't write about any obstacles doesn't necessarily mean they are naive and unaware of challenges. What we do know is that when these students responded, they did not focus on difficulties, and they seem "ready to go."

The realistic approach is also characterized by a great deal of enthusiasm and appreciation, even though these students also identify possible problems, quite likely to occur. For instance, both teachers and students need to be prepared. They also mention difficulties for the teacher to stay in charge. Despite the great challenge for a new teacher to put one's authority at risk, these students don't hesitate, and they seem very motivated to try drama in their own teaching. Their reflections are reasonable and can be understood as a sign of professional knowledge, being aware that it is not always easy to introduce something new in the classroom or school context. Nevertheless, their position seems to be "it may be difficult, but it is worth the effort."

The pessimistic approach to applying drama may be related to pre-service teaching experiences, perhaps colored by lack of flexibility, or other aspects like not feeling personally at ease with this kind of teaching. Another reason, or fear, is related to adolescent students in upper secondary school, who might find drama work embarrassing and more suitable for children. Yet another reason is connected to external demands, for instance, a strict headmaster. These students focus on hindrances and see very limited chances to overcome them. They seem to find it "very difficult to implement drama," and perhaps "not worth the trouble."

4.4.3 STUDENTS' REFLECTIONS ON SUSTAINABILITY AND THE ENVIRONMENT

Of the 56 students, nearly three-fourths (40 students) reflect on the theme of the drama workshop. In their responses, they refer explicitly to issues of sustainability, although most often in terms of environment or environmental problems. The students' responses connected to the theme have been organized into four areas, according to the varying focus of their reflections. These areas concern personal reflections, peer-related reflections, pedagogical reflections, and general reflections. In other words, reflections on sustainability are made at all levels, from the individual to this group of students, further to the future profession and finally regarding society as a whole. Below, these four areas are presented, mainly based on quotes.

Personal reflections: Some of these comments are brief and straightforward: "I learned about environmental problems." "I learned how to protect environment and to respect it." Others are a bit more thoughtful: "I realized that the environmental problems and the social problems become bigger as the years pass by." Yet others have a more affective tone: "It reminded me how much the environment needs us." To learn about what one can do is a first step, mentioned by several students: "I learned how to express my [concerns] for the environment and what I can do." To cooperate in small groups is also mentioned by several students: "I learned how to collaborate and what I can do about environment." The next step is, for instance, to actually change a habit. Some comments describe a willingness to act differently in one's own life: "It motivated me to do something in order to help environment protection."

I learned about environment protection and how I can change my life and do it more environmentally friendly.

4.4.3.1 Peer-Related Reflections

Many students reflect on the fact that they were asked to interact and discuss with each other. The workshop was seen as "an opportunity to [ex]change opinions and talk about real problems." "We did something 'out of the box' and we also learn important things about environment." Before the workshop, the students did not know much about each other's views, but "Now I know what my mates think." Comments like "I learn how other people see the environmental problems," "I realize other people care too about the environment," and "I liked that my colleagues care about environment" all point in this direction. Someone explicitly connects this to the drama work, "a chance to communicate of pollution and climate change and environment, thanks to the role-plays," "... and there was action. We also learn a lot about the environment and the social problems we have to face. We have to take action immediately."

"Through some special techniques, we became aware of the main problems our planet faces. We also got to know in an alternative way what's on my classmates mind."

4.4.3.2 Pedagogical Reflections

There are fewer comments within this area than in the others. Some of them are rather basic: "Good exercises are based on environmental problems," and "It's very important for the children because in this way [they] can understand better the environment problems." The workshop was considered to be "very helpful. It shows us how to teach alternatively about environment," while using drama "may challenge the students."

I learned many techniques I could use in environmental education. I have tried it once with a very good effect. I believe that drama can increase compassion for the earth to people that have not thought of it.

4.4.3.3 General Reflections

In this area, there are some comments involving "everyone" or "people." "Environmental sustainability is everybody's responsibility." It is seen as "a matter that is everyone's concern (or at least it should be)." "People should be informed about the bad results and what we could do to solve it." Some comments are more precise. "I learned things that I already know about environmental problems. But I also learn that all society levels can help to [support] the reduction." "Everyone is responsible for the protection of the environment. Not only government and companies but also citizens." Some especially brief comments are also placed under this heading, like "different stakeholder's group – different perspectives," or slightly more elaborated, "the significance of the engagement of all the parts in the facing of the environmental problems," Although these comments are made on a general level, it does not mean they are less sincere. "The most important thing I learned today is that all of us have to protect the environment in any way we can." The pollution of the environment has increased a lot, so all of us we should do something for that because we are all responsible for this situation.

4.4.4 COMMENTS ON STUDENTS' REFLECTIONS ON SUSTAINABILITY

From a personal perspective, the students focus on the importance of learning or finding out "what can I actually do." This is mentioned by several students, and connects to not only point out problems but also look at possible solutions (cf. Ojala, 2012b). Even if the effect of individual actions is very limited, these actions may serve to counteract passivity and hopelessness. And if many individuals make the same choices, it can definitely lead to change.

The significance of a shared experience, knowing "what my peers think about environment," was unexpected. It tells us that sustainability is frequently discussed in certain contexts, but may be not so often discussed among "ordinary" people. It seems like the students appreciated the chance to share their views and opinions regarding sustainability, by drama work, and to find out that they are not alone, but that their peers also care about the environment.

The pedagogical reflections are fairly general, and only two students wrote about using drama in teaching for sustainability. This is understandable, as the students still have very limited teaching experience and were completely unprepared to work on sustainability issues. Probably, this also explains why there are not so many comments about pedagogical aspects.

The students' reflections on sustainability at a general level, in terms of "everybody need to...," are not always clear. They can be understood as rather vague and withdrawn, as if "someone else" has to do "something," which could be a sign of a psychological defense strategy, trying to distance oneself from the problems. But it is also highly relevant to point out that we all need to make substantial adjustments of our lifestyle, and some students include themselves in "everyone," writing, for instance, "today I learned...that all of us..." The latter perspective seems to be the most common among these students.

The most interesting outcome from an ESD perspective is the responses that describe a motivation to act differently in one's own life.[3] Five students wrote, without being asked, about changing their own behavior to act more sustainable in their daily life. A handful of students wrote about *teaching* for sustainability, and some of them mentioned drama. The point here is not that they plan to apply drama, but that it seems likely for them to teach for sustainability. These students managed to grasp the whole idea of the workshop.

I find this very encouraging, after only one workshop. Of course, it is impossible to know if these changes actually happened, but the results indicate that in some cases the only thing needed is a little push (cf. nudging). This role-play was earlier conducted in Helsinki. In a follow-up enquiry, 1 year later, the Finnish teachers/postgraduate students remembered the workshop well, and some of them had made changes in their daily life (Österlind, 2018). This is in line with Buchs and Blanchard (2011) who reported that after a role-play, more than 90% of the students said it led to changed behavior – and this role-play was also a single event.

4.4.5   OVERVIEW OF THE RESULTS

A large majority (more than three-fourths of the students) clearly appreciate drama work as part of teaching and learning, although the students describe three different approaches to applying drama. One approach is very positive or enthusiastic. Of these 27 students, nobody mention any kind of pedagogical challenge or hesitation connected to using drama in their own teaching. Another approach is similarly positive to drama, but each of these 17 students makes some pedagogical reflections about possible difficulties when applying drama. The third approach is positive towards drama, but concerned about problems and obstacles. According to the comments of these 12 students, they are quite pessimistic, and perhaps less motivated, regarding the possibilities to implement drama in their own teaching. An overview of the data material and how it was structured is presented in Table 4.2.

4.4.6   APPRECIATION AND CRITIQUE OF THE DRAMA WORKSHOP

Independent of the students' approach to drama, or if they refer to the theme or not, there are some strongly positive comments and some precise criticism of the workshop. The students' responses related to the drama workshop as such are to a large extent highly appreciative: "I believe that [drama] is a clever way to make students think about a problem," and "a way to do our teaching more attractive." Some comments even refer to an overwhelmingly positive experience: "awesome idea, useful," and "playing role was fascinating," "extraordinary experience – remake it!"

---

[3] This was an explicit dimension of the workshop, but not mentioned in the questionnaire.

TABLE 4.2

Students' Responses: Dimensions in the Data Material

Approaches / Focus	Reflections on Drama	Reflections on Drama and the Environment	No
Positive to drama No problems mentioned	7 students	20 students	27
Positive to drama Some problems to handle	6 students	11 students	17
"Positive" to drama Severe problems foreseen	3 students	9 students	12
Total	**16 students**	**40 students**	**56**

However, there is also some specific critique of this particular workshop, regardless of the approach to drama. Three students criticize the format, as the workshop was going on for too long time (3–4 h): "It was a bit tiring in the end..." One of them suggests two shorter workshops would be a better solution. Another student finds the workshop nice and interesting, but "a little bit out of the purpose of the [study] program... I don't think that drama teaching is a good way to teach adults." Someone also mentions a lack of preparation, "If I were prepared for it, it would be fun." The most elaborated criticism is related to the content of the workshop. Obviously, such comments are highly interesting from a pedagogical perspective, and valuable in order to improve the intervention. Thus, I think it deserves to be presented in full detail.

It was really interesting but my problem was that no opinions were fully presented or developed for the others to have a better understanding of different points of view. It stayed in more superficial approaches and opinions that we all know (more or less). I saw the variety of opinions on problem-solving and the support for individual action with which I agree under very specific constraints. Drama constitutes a great teaching tool if used properly. It needs a good student preparation, specific instructive goals, and means. And of course, it needs good teachers' preparation and experience so as to reach the goals.

This comment highlights a dilemma between making it accessible to be part of a single role-play in a drama context, without any previous drama experience or preparation like readings, and avoiding stereotypical action, locked positions in in-role discussions, and so forth.

4.4.7 COMMENTS ON THE WORKSHOP AND THE STUDY

The critique about the workshop being too long and exhausting is of course relevant, also due to the fact that the work was conducted in a foreign language. The extended time had to do with the group size. Nearly 60 participants is a large number for a drama workshop, and it affects the whole teaching situation. Instead, 30 students would be preferable.

Regarding the elaborated critique of the role-play, I agree that it is more than likely the characters become one-sided, or even parodic, as very little time and emphasis is placed on the development of these on-spot created roles. The discussions also tend to be more or less superficial, depending on the students' varying knowledge of the content. In this case, the students were not at all prepared to discuss environmental problems and global warming from different perspectives. On the other hand, an unprepared role-play gives the teacher some information about what the students already know, and the students become aware of things they need to learn more about. This is confirmed by several comments about the need for students to be prepared. Thus, the role-play can serve as a starting point for teaching on sustainability. Students may also get a glimpse of why climate negotiations are so hard.

The intention was to give the participating students an introduction to educational drama work, and at the same time generate reflections and discussions about sustainability. To some extent, this was achieved, according to the data. The most encouraging comments after the workshop from an ESD perspective are made by the student-teachers who refer to their own teaching for sustainability in the future, and reflections that show an increased motivation and readiness to take action at a personal level.

The most evident limitation of this small-scale study is that (a) it concerns a single workshop and (b) there is no follow-up data. The first problem is out of reach for the teacher/researcher, but the second problem is definitely possible to address (Österlind, 2018). Follow-up studies are less demanding than longitudinal studies, even though both are needed. The lack of such studies has already been identified (Chen & Martin, 2015). An interesting initiative is a model developed to describe progression in learning for sustainability (see Greig & Priddle, 2019). Obviously, such a model is not designed to be applied when the teaching takes place on a single occasion.

There are considerable differences between a more cognitive, "academic" role-play, based on previous reading and careful student preparation, and a more affective, "ad hoc" role-play, based on the participants' personal thoughts and experiences. These differences make it less meaningful to compare role-plays as they were more or less similar. Many studies provide very little information about how the role-play was designed and carried out, which makes it even more difficult to compare the outcomes. I believe that role-playing in a drama context contributes to ESD by adding the affective dimension, as knowing facts is not enough (Ripple et al., 2017). And just like Forgazs (2013) and Lehtonen et al. (2018), I think that empathy and connectedness is a strong driving force for action.

4.5   CONCLUDING REMARKS

According to the data, the students seem to have learnt a lot – for instance, about themselves, about teaching and learning in general, and about drama as part of their teaching repertoire. A large majority (44 of the 56 respondents) reflect on both drama work and sustainability. We do not know if the positive impact directly after the intervention lasted over time, and in some cases led to changed behavior. This is of vital interest, and therefore, longitudinal research and/or follow-up studies are much needed in the field of ESD.

It might seem very tiny to offer the students one single workshop on a very complex topic like sustainability. But as research indicates, it may still have an impact. In this case, the students' written comments clearly indicate they did learn something about sustainability, and about drama work as well. These student-teachers made a lot of connections to their own future teaching, and some of them explicitly included teaching for sustainability.

From the perspective of SDGs, and teaching for sustainability in HE, the most promising comments directly after the workshop are made by the student-teachers who refer to their own teaching for sustainability in the future, with or without elements of drama, and reflections that indicate an increased motivation and readiness to take action at a personal level. This overall comment quite beautifully describes.

We tried to understand environmental problems, find the causes, and suggest solutions while searching inside of us, using our own opinions, sensitivity, and experiences. Actually, I understood that I'm able to do things about the society and that I can be the beginning and the changing of what I want to see in the world.

As education, in general, is seen as a key factor for sustainable development, teacher education becomes a highly significant field, holding a great potential. It would be much easier if sustainability was treated as any academic content with allocated teaching time and resources. But the strategy to (slowly) let the concept and perspective of sustainability permeate all kinds of HE might turn out to have a greater impact in the long run.

REFERENCES

Álvarez-García, O., García-Escudero, L. A., Salvà-Mut, F., and Calvo-Sastre, A. (2019). Variables influencing pre-service teacher training in education for sustainable development: A case study of two Spanish Universities. *Sustainability* 11, 4412.

Ballantyne, R. and Packer, J. (2007). Introducing a fifth pedagogy: Experience-based strategies for facilitating learning in natural environments. *Environmental Education Research* 15(2), 243–262. doi: 10.1080/13504620802711282.

Björnberg, K. E., Skogh, I.-B., and Strömberg, E. (2015). Integrating social sustainability in engineering education at the KTH Royal Institute of Technology. *International Journal of Sustainability in Higher Education* 16(5), 639–649. doi: 10.1108/IJSHE-01-2014-0010.

Blanchard, O. and Buchs, A. (2015). Clarifying sustainable development concepts through role-play. *Simulation and Gaming*. doi: 10.1177/1046878114564508.

van Boeckel, J. (2009). Arts-based environmental education and the ecological crisis: Between opening the senses and coping with psychic numbing. In: B. Drillsma-Milgrom & L. Kirstinä (Eds.) *Metamorphoses in Children's Literature and Culture* (pp. 145–164). Turku: Enostone.

van Boeckel, J. (2013). *At the Heart of Art and Earth: An Exploration of Practices in Arts-Based Environmental Education*. Helsinki: Aalto University.

Bogdan, R. C. and Biklen, S. K. (2003). *Qualitative Research for Education: An Introduction to Theories and Methods*. Boston, MA: Allyn and Bacon.

Buchs, A. and Blanchard, O. (2011). Exploring the concept of sustainable development through role-playing. *Journal of Economic Education* 42, 388–394.

Cantell, H., Tolppanen, S., Aarnio-Linnanvuori, E., and Lehtonen, A. (2019). Bicycle model on climate change education: Presenting and evaluating a model. *Environmental Education Research* 25(5), 1–15.

Chen, J. C. and Martin, A. R. (2015). Role-play simulations as a transformative methodology in environmental education. *Journal of Transformative Education* 13(1), 85–102. doi: 10.1177/1541344614560196.

Christie, B. A., Miller, K. K., Cooke, R., and White, J. G. (2013). Environmental sustainability in higher education: How do academics teach? *Environmental Education Research* 19(3), 385–414. doi: 10.1080/13504622.2012.698598.

Cruickshank, H. and Fenner, R. (2012). Exploring key sustainable development themes through learning activities. *International Journal of Sustainability in Higher Education* 13(3), 249–262.

Davis, S. and Tarrant, M. (2014) Environmentalism, stories and science: Exploring applied theatre processes for sustainability education. *Research in Drama Education: The Journal of Applied Theatre and Performance* 19(2), 190–194. doi: 10.1080/13569783.2014.895613.

Dewey, J. (1934). *Art as Experience.* New York: Perigee Books.

Díaz-Iso, A., Eizaguirre, A., and García-Olalla, A. (2019). Extracurricular activities in higher education and the promotion of reflective learning for sustainability. *Sustainability* 11, 4521.

Eernstman, N. and Wals, A. E. J. (2013). Locative meaning-making: An arts-based approach to learning for sustainable development. *Sustainability* 5, 1645–1660.

Eliason Bjurström, Å. (2012). Trying to make 'reality' appear different: Working with drama in an intercultural ESD setting. *Social Alternatives* 31(4), 18–23.

Elo, S., Kääriäinen, M., Kanste, O., Pölkki, T., Utriainen, K., and Kyngäs, H. (2014). Qualitative content analysis: A focus on trustworthiness. *SAGE Open* 4(1), 1–10.

Forgazs, R. (2013). Response to the themed issue: Environmentalism. *Research in Drama Education: The Journal of Applied Theatre and Performance* 18(3), 324–328. doi: 10.1080/13569783.2013.810928.

Glasser, H. (2018). Toward robust foundations for sustainable well-being societies: Learning to change by changing how we learn. In: J. Cook (Ed.) *Sustainability, Human Well-Being, and the Future of Education* (pp. 31–89). London: Palgrave Macmillan.

Gordon, S. and Thomas, I. (2016) 'The learning sticks': Reflections on a case study of role-playing for sustainability. *Environmental Education Research.* doi: 10.1080/13504622.2016.1190959.

Greig, A. and Priddle, J. (2019). Mapping students' development in response to sustainability education: A conceptual model. *Sustainability* 11, 4324.

Heras, M. and Tàbara, J. D. (2014). Let's play transformations! Performative methods for sustainability. *Sustainability Science* 9(3), 379–398.

Heyward, P. (2010). Emotional engagement through drama: Strategies to assist learning through role play. *International Journal of Teaching and Learning in Higher Education* 22(2), 197–203.

Hunter, M., Aprill, A., Hill, A., and Emery S. (2018). *Education, Arts and Sustainability: Emerging Practice for a Changing World.* Singapore: Springer.

Kvale, S. (2007). *Doing Interviews.* London: Sage.

Læssøe, J. (2010). Education for sustainable development, participation and socio-cultural change. *Environmental Education Research* 16(1), 39–57.

Laininen, E. (2018). Transforming our worldview towards a sustainable future. In: J. Cook (Ed.) *Learning at the Edge of History.* London: Palgrave, pp. 161–200.

Leal Filho, W., Shiel, C., Paço, A., Mifsud, M., Veiga Ávila, L., Londero Brandli, L., Molthan-Hill, P., Pace, P., Azeiteiro, U. M., Ruiz Vargas, V., and Caeiro, S. (2019). Sustainable Development Goals and sustainability teaching at universities: Falling behind or getting ahead of the pack? *Journal of Cleaner Production* 232, (285–294).

Lehtonen, A., Salonen, A., Cantell, H., and Riuttanen, L. (2018). A pedagogy of inter-connectedness for encountering climate change as a wicked sustainability problem. *Journal of Cleaner Production* 199, 860–867.

Lotz-Sisitka, H., Wals, A. E., Kronlid, D., and McGarry, D. (2015). Transformative, trans-gressive social learning: Rethinking higher education pedagogy in times of systemic global dysfunction. *Current Opinion in Environmental Sustainability* 16, 73–80.

Lundegård, I. and Wickman, P.-O. (2007). Conflicts of interest: An indispensable element of education for sustainable development. *Environmental Education Research* 13(1), 1–15.

Mezirow, J. (2000). *Learning as Transformation: Critical Perspectives on a Theory in Progress*. The Jossey-Bass Higher and Adult Education Series. San Francisco, CA: Jossey-Bass Publishers.

Ojala, M. (2012a). Regulating worry, promoting hope: How do children, adolescents and young adults cope with climate change? *International Journal of Environmental and Science Education* 7(4), 537–561.

Ojala, M. (2012b). Hope and climate change: The importance of hope for environmental engagement among young people. *Environmental Education Research* 18(5), 625–642.

Oliver, S. (2016). Integrating role-play with case study and carbon footprint monitoring: A Transformative approach to enhancing learners' social behavior for a more sustain-able environment. *International Journal of Environmental and Science Education* 11(6), 1323–1335. doi: 10.12973/ijese.2016.346a.

Österlind, E. (2008). Acting out of habits: Can theatre of the oppressed promote change? Boal's theatre methods in relation to Bourdieu's concept of Habitus. *Research in Drama Education* 13(1), 71–82.

Österlind, E. (2012). Emotions-aesthetics-education: Dilemmas related to students' com-mitment in education for sustainable development. *Journal of Artistic and Creative Education* 6(1), 32–50.

Österlind, E. (2018). Drama in higher education for sustainability: Work-based learning through fiction? *Higher Education, Skills and Work-Based Learning* 8(3), 337–352.

Paschall, M. and Wüstenhagen, R. (2012). More than a game: Learning about climate change through role play. *Journal of Management Education* 36(4), 510–543.

Roth, W. M. (2014). From within the event: A post constructivist perspective on activism, ethics, and science education. In: L. Bencze & S. Alsop (Eds.) *Activist Science and Technology Education* (pp. 237–254). Dordrecht: Springer.

Schnurr, M. A., De Santo, E. M., and Green, A. D. (2014). What do students learn from a role-play simulation of an international negotiation? *Journal of Geography in Higher Education* 38(3), 401–414. doi: 10.1080/03098265.2014.933789.

Tooth, R. and Renshaw, P. (2009). Reflections on pedagogy and place: A journey into learn-ing for sustainability through environmental narrative and deep attentive reflection. *Australian Journal of Environmental Education* 25, 95–104.

UNESCO. (2005). *The DESD at a Glance*. Paris: UNESCO. http://unesdoc.unesco.org/images/0014/001416/141629e.pdf.

UNESCO. (2010). Seoul Agenda: Goals for the Development of Arts Education. UNESCO.

UNESCO. (2016). Three challenges for higher education and the SDGs. www.iiep.unesco.org/en/three-challenges-higher-education-and-sdgs-3556 (downloaded 2018-03-26).

UNESCO. (2017). Education for sustainable development goals: Learning objectives. Paris: United Nations Educational, Scientific and Cultural Organization. https://unesdoc.unesco.org/ark:/48223/pf0000247444. Retrieved February 14 2019.

Wall, T. and Hindley, A. (2018). Work integrated learning for sustainability education. In: W. Leal Filho et al. (Eds.) *Encyclopedia of Sustainability in Higher Education*. Dordrecht: Springer, pp. 226–232.

Wall, T., Österlind, E., and Fries, J. (2018a). Art-based teaching on sustainable development. In: W. L. Filho (Ed.) *Encyclopedia of Sustainability in Higher Education.* New York: Springer Nature, pp. 1–7.

Wall, T., Österlind, E., and Fries, J. (2018b). Arts based approaches for sustainability. In: W. L. Filho (Ed.) *Encyclopedia of Sustainability in Higher Education.* New York: Springer Nature, pp. 1–7.

Wareham, D. G., Elefsiniotis, T. P., and Elms, D. G. (2006). Introducing ethics using structured controversies. *European Journal of Engineering Education* 31(6), 651–660. doi: 10.1080/030437906009t.

5 Knowledge Creation and Sharing in Higher Education
The Long-Run Effect in Sustainable Development

Athanassios Androutsos
Athens University of Economics and Business

CONTENTS

Scientia est potentia
(Knowledge is power)

Sir Francis Bacon
Meditationes Sacrae, 1597

5.1 INTRODUCTION

One of the main factors related to sustainable development is knowledge (Richardson and Dyball, 2016). Knowledge refers to the information capacity that each human encodes in his head. Knowledge, including skills and competences, is at the core of the education system. From the dawn of the 21st century since today, the global economy has been going through a great transformation; the fourth industrial revolution (Industry 4.0) (Vuksanović et al., 2016), Globalization 4.0 (World Economic

Forum, 2019), and Society 5.0 (Harayama, 2017) have introduced a new economic culture based on technological innovation and human-centered society. In Society 5.0, new wealth will be created through innovation. Economic growth will be constantly increasing, and at the same time, solutions will be given to social problems. A prerequisite for this to happen is that the quality of the human capital that mainly comes from the higher education system is high. The 21st-century skills that are needed in future jobs are closely related to innovation and creativity, and thus, creating new knowledge is a fundamental competency and goal of the current and future education system (Androutsos and Brinia, 2019). Consequently, the students of the current higher education system need to be transformed from passive receivers and consumers or even explorers of information into participants and creators of new knowledge (Tahir, 2013; Kurniawan, 2014). Moreover, the gap between academia and entrepreneurship needs to get shrunk down and students have to collaborate with enterprises in various types of projects (Hasan et al., 2017). Students and graduates of higher education have fresh ideas that could be transformed into products and services by enterprises (Roncaglia, 2005).

Sustainable development could be defined as the state of the economy where growth rates are not decreasing in the long run due to the exploitation of natural resources but remain constant in the steady state (Solow, 1974; Heal, 1996; Chichilnisky, 1997; Mittal and Gupta, 2015). Also, optimal sustainable development is balanced; that is, consumption equals savings in the long run. This is the golden rule of sustainable development and is dependent on how much we value consumption in the present over future generations (Phelps, 1961). Knowledge as a production factor may generate sustained and balanced growth rates and thus sustainable development in the long run since knowledge is unlimited, non-rival, and nonexclusive (Romer, 1990; Economides and Philippopoulos, 2007; Enachi, 2009; OECD 2010).

Since knowledge is a type of information, it has a unique feature: it can be reproduced and shared at almost zero marginal cost (Koski, 2011). Zero marginal cost sharing of information and knowledge is closely related to the idea of positive knowledge externalities or spillovers (Buchanan and Stubblebine, 1962): each student's decision on creating new types of knowledge affects all other students' level of knowledge through sharing. In this paper, we investigate the effect of knowledge creation and sharing in higher education economy in order to generate sustained and balanced economic growth rates, that is, to generate sustainable development.

In terms of a production function, we can think of knowledge as a measure of the information in each student's brain together with the skills, competences, and values of the students in higher education. Information sharing, on the other hand, could be considered as a measure of the utility that students receive from consuming other students' knowledge and designs. The utility comes in the form of a virtual flow of information from one person's brain to the other person's brain. This utility, then, raises the ability of the receiver to be more productive in terms of newly created knowledge. Since information sharing is not excludable and could be delivered at almost zero marginal cost, it justifies knowledge externalities or spillovers (Romer, 1990).

In this paper, we propose the idea of a higher education economy that is structurally similar to start-ups economy. In other words, we consider a market where enterprises will purchase knowledge-based designs from the students and transform them into products or services. Moreover, our economy will share its accumulated knowledge through a social-like communication network that will provide connection services to the members of our economy and give them the opportunity to share and also to exploit the accumulated stock of knowledge.

5.2   CONCEPTUAL FRAMEWORK AND LITERATURE REVIEW

For better understanding the field of economic growth, we can consider a Robinson Crusoe-like economic setup (Mathias, 2007). Robinson Crusoe owns the physical capital of his own economy such as fishing weapons and other machinery or technology and transforms these inputs into outputs such as fishes (i.e., consumable goods) and new weapons (i.e., capital goods). R. Crusoe must decide how much time to spend on fishing and thus increasing his consumption today and how much time to spend on investing in new technologies (e.g., new weapons) for increasing his consumption in the future. In order to maximize his consumption in the long run, Robinson Crusoe needs to compare the benefits and costs of consuming today rather than tomorrow.

Expanding R. Crusoe's model, we assume that the only asset in our higher education economy is H_t. H_t includes several inputs that can be accumulated such as knowledge and skills. Let us assume that the production function of our higher education economy has the following form:

$$Y_t = F(H_t, S_t) \qquad (5.1)$$

where knowledge H_t could be accumulated and the other, the population of our economy, S_t, cannot be accumulated; that is, its growth rate does not depend on individual choices. You may want to think of S_t as students but it may also include other members of the higher education system, such as alumni that work in enterprises. Y_t is the output, that is, the designs and knowledge-based products produced by students at time t.

The students decide how much time to spend on consuming knowledge or investing their time in creating new knowledge and new designs. If they invest more time in creating and accumulating new knowledge, students will have less time to spend in information consumption and hence their consumption in the present will be decreased and therefore their consumption in the future will be increased. The net increase in students' knowledge H_t over time, $\dot{H}_t = dH_t/dt$, is the aggregate newly created knowledge that could be seen as an investment in our economy, where a dot over $\dot{H}_t$ denotes differentiation with respect to continuous time. Hence, knowledge H_t grows according to the following functional form:

$$\dot{H}_t = sF(H_t, S_t) = F(H_t, S_t) - C_t \qquad (5.2)$$

where $s \in [0,1]$ is the saving rate of output $F(H_t, S_t)$, that is, the fraction of time that is assigned in creating new knowledge. $\dot{H}_t$ can be seen alternatively as a function of forgone consumption or as a function of knowledge accumulation.

We define lower case h_t, be the knowledge-student-population ratio or knowledge per capita $h_t = H_t/S_t$ at time t and c_t be the consumption per capita $c_t = C_t/S_t$. Assuming that S_t does not grow (i.e., $\dot{S}/S = 0$), we have that knowledge per capita is growing according to the following functional form:

$$\dot{h}_t = F(h_t) - c_t \tag{5.3}$$

At the long-run optimal equilibrium, all variables (i.e., knowledge consumption and output) grow at a constant rate. This state of optimal growth rates could be considered as a balanced and sustained state where the economy grows with optimal growth rates in the long run.

In Solow and Swan's growth model (Solow, 1956; Swan, 1956), it is assumed a Cobb-Douglas production function (Saini, 1974):

$$Y_t = AK_t^\beta L_t^\alpha \tag{5.4}$$

where A denotes the technology for transforming inputs into output. Since we assume that $\alpha + \beta = 1$, the functional form implies that the production function exhibits constant returns of scale (CRS), but, since $\alpha < 1$, $\beta < 1$, it exhibits decreasing returns to K_t and L_t alone. Solow-Swan's growth model is not optimal in the long run since the saving rate is a choice variable and not the outcome of an optimization process.

Cass and Koopmans's growth model (Cass, 1965; Koopmans, 1965) and Ramsey's analysis of consumer optimization (Ramsey, 1928) allowed for the optimal determination of the consumption path since the saving rate of the economy, that is, the investment rate, was the outcome of the optimization process. In Cass and Koopmans's growth model, it is assumed that consumers maximize their utility in the long run by choosing a consumption path that optimizes an intertemporal utility function, $u(c_t)$, subject to intertemporal constraints. The utility function takes the form of Ramsey's constant intertemporal elasticity of substitution utility function:

$$U(0) = \int_0^\infty e^{-\rho t} u(c_t)\,dt = \int_0^\infty e^{-\rho t}\left(\frac{c_t^{1-\sigma} - 1}{1 - \sigma}\right) dt \tag{5.5}$$

where ρ is the intertemporal discount rate, which discounts future consumption in the present, and it could be interpreted as that students could value more their own consumption utility than next generations' utility and σ is the intertemporal elasticity of substitution. The intertemporal elasticity of substitution is the intention to consume today as if there was no tomorrow.

The long-run growth rate at steady state of both Solow-Swan and Ramsey-Cass-Koopmans growth models is zero, since $\beta < 1$, and the marginal product of K_t is constantly decreasing, thus resulting in zero growth rates.

In order to have positive growth rates at the long-run equilibrium, the production function should be assumed to have constant returns with regard to the inputs that can be accumulated:

$$Y_t = AK_t \tag{5.6}$$

This approach is taken by Rebelo (1990) where K_t is a broad measure of inputs that can be accumulated. Lucas (1988) and Uzawa (1965) also followed the same approach. In these models, the optimal growth rates are determined without exogenous constraints, and thus, these types of models are considered endogenous growth models.

In other cases (Barro and Sala-i-Martin, 2004), the CRS production function could be combined with positive production externalities (i.e., positive side effects of the models) that increase the production functions to IRS (increasing returns of scale). These models are often called IRS endogenous growth models.

However, the long-run equilibrium in those models is not optimal (Varian, 1992). In order to address this issue, the externalities could be internalized. In other words, each economic agent affects all other agents' output but none of the individual agents takes this side effect into account, since each agent thinks he is too small to affect aggregate variables. However, the economy at the aggregate level takes externalities into account and assumes an IRS production function that creates greater growth rates. This approach will be followed in our model as well.

In pedagogical terms, the proposed model is the economic interpretation of design-thinking pedagogy (Androutsos and Brinia, 2019; Luka, 2014). Design thinking is a business paradigm that promotes innovation and entrepreneurship based on design-erly ways of thinking and making (Brown, 2009). When design thinking is applied to higher education, it leads to innovative results and newly created knowledge by the students. Design thinking is an open process that is applied inside and outside the school doors in collaboration with enterprises (Huber et al., 2016). Entrepreneurship is a crucial dimension of design thinking that gives the ability to the students to implement their ideas and designs in collaboration with enterprises and transform user preferences into demand. Thus, the proposed model is the economic conceptualization of design-thinking pedagogy. Higher education knowledge-based economy and design-thinking pedagogy are two sides of the same coin.

In order to have a more realistic view of the economy that will be proposed and get described, we could think of the start-up ecosystem that is structurally similar to the proposed higher education economy (Salamzadeh and Kawamorita, 2017). Start-ups are mainly consisting of higher education people, collaborate with people from academia and thus create new knowledge and transform this knowledge to new designs and prototypes. The collaborative enterprises and angel investors purchase knowledge designs or invest and finance the start-ups (KPMG, 2019).

5.3 METHODOLOGY

We assume a knowledge-based higher education economy where the only input is knowledge. The goods that are produced and sold are designs produced by the students and delivered to enterprises. Enterprises purchase knowledge-based products by the students (knowledge is assumed to be owned by the students) and use those designs to produce products and services.

Students have access to learning methods such as design-thinking pedagogy and other technology and techniques such as digital tools, art-based practices, and entrepreneurial methods that allow them to transform knowledge into designs.

We also consider a communication network over the web, which is a special type of social network that gives the ability to the students, graduates, and entrepreneurial people to connect to each other and share their knowledge and designs. We assume that knowledge-based output is protected by intellectual property rights and public copyright licenses that enable the free distribution and sharing of the otherwise copyrighted output.

Students are maximizing their consumption utility by choosing the fractions of time to assign on consumption and knowledge creation, respectively. Hence, students choose a knowledge consumption path in order to maximize their intertemporal utility function.

5.3.1 KNOWLEDGE-BASED INTERTEMPORAL UTILITY FUNCTION

We assume a utility function of the following Ramsey's form:

$$U(0) = \int_0^\infty e^{-\rho t} u(c_t) dt = \int_0^\infty e^{-\rho t} \left(\frac{c_t^{1-\sigma} - 1}{1 - \sigma} \right) dt \qquad (5.7)$$

This functional form implies that the students of our higher education economy choose a consumption path of information so as to maximize their long-term utility. We define $U(0)$ as the present-valued integral of all the present and future knowledge-based utilities. c_t is the knowledge consumption per student, and $u(c_t)$ is the instantaneous per student happiness from the consumption of knowledge. Thus, $e^{-\rho t} u(c_t)$ is the instantaneous utility discounted at the present, whereas ρ is the discount rate which represents how much we value knowledge-based services and sustainable development over future generations. Parameter σ is the intertemporal elasticity of substitution between knowledge-based consumption and savings, that is, the willingness to intertemporally substitute knowledge consumption.

5.3.2 SHARING EFFECTS

Assume a random student n in the higher education-sharing network. The overall knowledge-based output, y_n, created by student n will be a function of his own knowledge h_n and the fraction of knowledge $\tilde{h}$ that we assume that the student happened to find in the network (i.e., the sharing effect). We assume that the sharing effect $\tilde{h}$ has a probability distribution $1/|N|$, where N is the student population. Thus, we assume that the expected value or mean of the sharing effect $\tilde{h}$ is equal to: $\tilde{h} = \dfrac{1}{|N|} \sum_{i=1}^{|N|} n_i$.

That is, the sharing effect equals to average per student knowledge. Consequently, the knowledge-based production function takes the form:

$$y_r = f\left(h_n, \tilde{h}\right) \qquad (5.8)$$

We consider knowledge h_n as a broad measure of knowledge and skills of the student. It is apparent that knowledge is a cumulative input that exhibits constant returns of scale with respect to knowledge-based output. In other words, if we multiply knowledge h_n by $\lambda > 1$, we will get λ times as much knowledge-based output. Thus, we have $y_n = f(h_n)$. On the other hand, if we increase the sharing effect, that is, average knowledge, λ times, it follows that knowledge-based output will be increased less than λ times. Since average knowledge is increasing, it is getting more difficult to get internalized by the students, and thus, average knowledge exhibits decreasing returns to scale with respect to knowledge-based output. Thus, we have $y_n = f(\tilde{h}^\phi)$, where ϕ is some positive constant and $\phi < 1$. Finally, the higher education economy production function will have the following form in per capita (per student) terms:

$$y_n = Ah_n \tilde{h}^\phi \tag{5.9}$$

where $\phi < 1$ denotes $\tilde{h}$ elasticity with respect to knowledge-based output per student, and A is the methodology (i.e., design thinking) used to transform knowledge into knowledge-based products.

5.3.3 Sharing and Knowledge Externalities

An important assumption of this model is that since the information and knowledge stock of each individual student is shared through a communication network, any increase in each individual student's knowledge increases all other students' knowledge-based output since it increases average knowledge $\tilde{h}\left(\tilde{h} = \sum_{i=1}^{|N|} h_i / |N|\right)$ of the community. This effect implies positive knowledge externalities: each student's increment on his own knowledge level increases all other students' output, since shared information becomes an input for each student without a direct cost.

Assume a representative student n which increases his knowledge volume. According to Eq. (5.9), since average knowledge of students' community is increased, there is an additional benefit for each individual student.

In a higher education economy without a central administrative authority for knowledge management, each individual student thinks he is too small to affect aggregate variables (i.e., aggregate and average knowledge), and thus, students do not take knowledge externalities into consideration; hence, students think that average knowledge is a constant factor and take it as given.

At the aggregate level, however, higher education economy takes knowledge externalities into consideration, and thus, $\tilde{h}$ (i.e., average knowledge) is not assumed as being given. In higher education economy as a whole which is managed by a single administrative authority or "social planner," knowledge externalities are internalized and average knowledge is considered identical to each individual student's knowledge. Therefore, $\tilde{h} \equiv h$. Thus, in higher education economy as a whole, the production function is as follows:

$$y_n = Ah_n^{1+\phi} \tag{5.10}$$

5.3.3.1 Individual Student's Economy

Is this model students choose a knowledge consumption path for maximizing their intertemporal utility. The maximization problem is as follows:

$$MAXU(0) = \int_0^\infty e^{-\rho t}\left(\frac{c_t^{1-\sigma}-1}{1-\sigma}\right)dt \qquad (5.11)$$

such that

$$\dot{y}_t = Ah_t\tilde{h}_t^\phi - c_t \qquad (5.12)$$

$$h_0 > 0 \qquad (5.13)$$

$$0 \le c_t \le y_t \qquad (5.14)$$

By solving the above problem, we get the equation for knowledge maximization in the long run:

$$\rho + \sigma\left(\dot{c}_t/c_t\right) = Ah_t^\phi \qquad (5.15)$$

The left-hand side of Eq. (5.15), $\rho + \sigma\left(\dot{c}_t/c_t\right)$, represents the instantaneous return to knowledge-based consumption, whereas the right-hand side, Ah_t^ϕ, is the instantaneous return to investment in knowledge. Based on Eq. (5.15), we get the growth rate of knowledge consumption, γ_c:

$$\gamma_c = \left(\dot{c}_t/c_t\right) = \left(Ah_t^\phi - \rho\right)/\sigma \qquad (5.16)$$

For γ_c being positive at $t = 0$, it should be: $Ah_0^\phi - \rho > 0 \Rightarrow h_0 > (\rho/A)^{1/\phi}$. Higher education economies that are regional or national economies with initial per student knowledge stock less than $(\rho/A)^{1/\phi}$ will get into a poverty trap; that is, there is no adequate knowledge stock for growth to get started. Moreover, based on Eq. (5.16), γ_c, the first-order condition of γ_c with respect to h_t is $A\phi h_t^{\phi-1}/\sigma > 0$, which asymptotically goes to zero. Thus, there is a point in time, in the long run, where we can assume that $h_t = h^*$ and where knowledge growth rate will become constant, as depicted in Figure 5.1.

Hence, in steady state, the growth rate will be $g^* = \left(Ah^{*\phi} - \rho\right)/\sigma$.

5.3.3.2 Higher Education Economy as a Whole

At the level of higher education aggregate economy, we internalize knowledge externalities. We consider a "social planner" (e.g., a strategic public authority) who understands that average knowledge is affected by individual student decisions, and hence, he assumes $\tilde{h} \equiv h$. Therefore, he internalizes knowledge externalities by

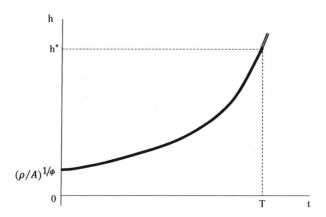

FIGURE 5.1   Knowledge per student growth in a competitive higher education economy.

setting $y_t = Ah_t^{1+\phi}$ as the production function of the economy, and he considers the following problem:

$$MAX U(0) = \int_0^\infty e^{-\rho t}\left(\frac{c_t^{1-\sigma}-1}{1-\sigma}\right)dt \qquad (5.17)$$

such that

$$\dot{y}_t = Ah_t^{1+\phi} - c_t \qquad (5.18)$$

$$h_0 > 0 \qquad (5.19)$$

$$0 \le c_t \le y_t \qquad (5.20)$$

If we solve the above problem, we get the equation for knowledge consumption maximization in the aggregate level:

$$\rho + \sigma\left(\dot{c}_t/c_t\right) = A(1+\varphi)h_t^\phi \qquad (5.21)$$

The left-hand side of Eq. (5.21), $\rho + \sigma\left(\dot{c}_t/c_t\right)$, denotes the instantaneous return to knowledge design consumption, while the right-hand side, $A(1+\varphi)h_t^\phi$, denotes the instantaneous return to investment in knowledge. Assume that γ_{planner} is the growth rate in the aggregate higher education economy. Then, the social planner's growth rate is given by the following equation:

$$\gamma_{\text{planner}} = \left[A(1+\phi)h_t^{*\phi} - \rho\right]/\sigma \qquad (5.22)$$

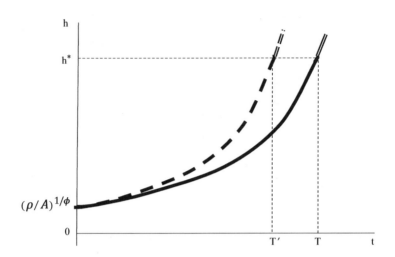

FIGURE 5.2   Social planner (dashed line) vs. competitive economy (solid line).

For having $\gamma_{\text{planner}} > 0$ at the beginning of time ($t = 0$), we should have:

$$A(1+\phi)h_0^\phi - \rho > 0 \Rightarrow h_0 > \left\{\rho / \left[A(1+\phi)\right]\right\}^{1/\phi} \tag{5.23}$$

Thus, economies with initial per student knowledge less than $\left\{\rho / \left[A(1+\phi)\right]\right\}^{1/\phi}$ will get into a poverty trap. Alike individual student's economy, the aggregate economy will reach constant growth rates at a steady state.

However, due to knowledge externalities, the aggregate student economy grows faster and consequently reaches a steady state at time $T' < T$. Figure 5.2 depicts aggregate growth rate vs. individual student's growth rate.

Individual student's higher education economy does not grow enough. The reason is that the private return to investment in knowledge is lower than the social one due to knowledge externalities. Hence, the integration of higher education institutions in an aggregate community and economy managed by a single strategic administrator will further increase the rate of growth.

5.4   RESULTS AND DISCUSSION

A structurally similar economy to what has been referred to as higher education economy is the start-up ecosystem. Basically, the start-up scene could be considered as a higher education economy since there is a close collaboration between academia and the private sector while the great majority of the people involved are coming from the higher education system. According to EU Startup Monitor report (Steigertahl and Mauer, 2018), more than 94% of start-up founders are coming from the higher education system. Moreover, there is a market for knowledge-based products. Thus, we can assume that the start-up scene is a type of higher education economy with similar parameters A, ρ, σ, and ϕ.

Assuming that the researchers' population growth rate is zero, knowledge growth in the start-up economy should be attributed to knowledge designs per capita growth that are produced by the researchers and "consumed" by the investors. Thus, the objective of our accounting methodology will be to investigate the investment's growth rate in the start-up scene, which, according to our model, will be equal to per capita knowledge accumulation.

According to Finish Business Angels Network (FiBAN, 2018), financing early-stage growth companies in Finland is growing in an exponential rate. As expected by the proposed model and confirmed by data (Figure 5.3), the shape of the curve is exponential.

According to Moody's Analytics (Moody's, 2015), the estimated Silicon Valley growth index is growing exponentially (Figure 5.4). Silicon Valley economy is a

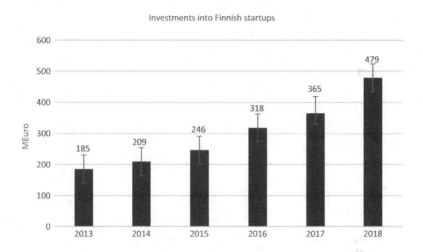

FIGURE 5.3   Financing early-stage growth companies in Finland.

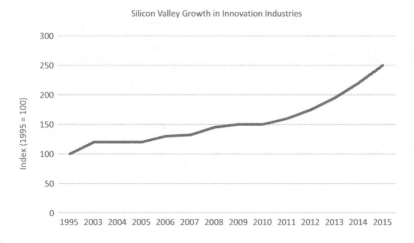

FIGURE 5.4   Silicon Valley growth index.

knowledge-based and knowledge-sensitive economy. The blue-colored curve represents the growth rate of the growth index, which also exhibits the exponential growth rates and confirms the proposed model.

In Africa, according to a report of the Partech Africa Team, 2018 was a monumental year in the start-up scene (Collon and Dème, 2018). The funding raised by African tech start-ups exhibited a growth rate of 33% in 2016, 53% in 2017, and 108% in 2018 (Figure 5.5). This represents an exponential growth rate that also confirms the prediction of the proposed model.

Also, in an analysis of OECD using data from Crunchbase (see Figure 5.6), it is found that the growth rate in the global artificial intelligence start-up scene is exponential (OECD, 2018).

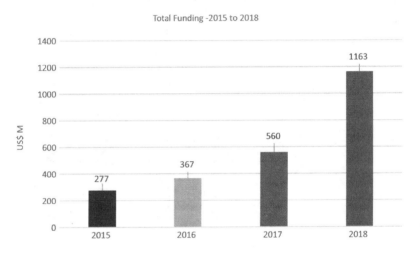

FIGURE 5.5   African tech start-ups total funding: 2015–2018 (in US$ M).

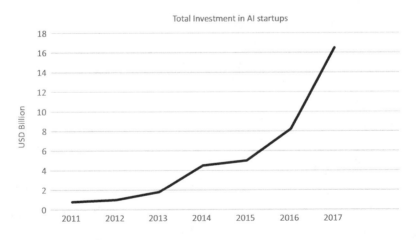

FIGURE 5.6   Private equity investment in artificial intelligence.

It is, hence, apparent that there is a relation between start-up economies and exponential growth rates, which confirms the predictions of the proposed model.

5.5 CONCLUDING REMARKS

We have presented an approach to optimal, balanced, and sustained economic growth in the long run based on knowledge externalities created in the higher education economy. The proposed model predicts exponential growth rate of the stock of knowledge created in higher education system when there is a market for knowledge designs. It also predicts differences in growth rates between countries or regions based on the effectiveness of the higher education system to transform knowledge into designs and demand. Several data sets and diagrams have been compiled to explore the empirical knowledge growth rates, which can be measured as the amount of investment in a structurally similar economy such as the start-ups economy. The findings confirm the results of the proposed model.

The results of our work are essential for better understanding the implications of higher education knowledge creation and sharing in sustainable growth. However, there are issues that need further investigation both in the pedagogical and in the economic field of research. From a pedagogical perspective, the higher education system needs a new pedagogy that will promote innovation, creativity, and entrepreneurship and will provide the appropriate skills and competencies to graduates to be able to create sustainable wealth. From an economic perspective, we need to create and strengthen collaborative opportunities between universities and enterprises by creating an education ecosystem and market for knowledge that will be similar to the start-up market. We also need a new type of social networking between students, graduates, and entrepreneurial people in order to share knowledge and thus increase the rate of knowledge and economic growth. The effect of interconnection between knowledge creators inside and outside of the higher education system eventually determines the overall economic growth and consequently affects sustainability and people's prosperity in the 21st century.

REFERENCES

Androutsos, A. and Brinia, V. (2019). Developing and piloting a pedagogy for teaching innovation, collaboration, and co-creation in secondary education based on design thinking, digital transformation, and entrepreneurship. *Education Sciences*, 9, 113.

Barro, R. J. and Sala-i-Martin, X. (2004). *Economic Growth*, Second Edition. London, England: The MIT Press.

Brown, T. (2009) *Change by Design: How Design Thinking Transforms Organizations and Inspires Innovation*. New York: HarperBusines.

Buchanan, J. and Stubblebine, C. (1962). Externality. *Economica*, 29(116), 371–384. doi: 10.2307/2551386.

Cass, D. (1965). Optimum growth in an aggregate model of capital accumulation. *Review of Economic Studies*, 32, 233–240.

Chichilnisky, G. (1997). What is sustainable development. Available at SSRN: https://ssrn.com/abstract=1375216 or http://dx.doi.org/10.2139/ssrn.1375216.

Collon, C. and Dème, T. (2018). 2018 was a Monumental Year for African Tech Start-ups, with US$ 1.163 Billion raised in equity funding, a 108% YoY Growth. Report by the Partech Africa Team. Partech.

Economides, G. and Philippopoulos, A. (2007). Growth enhancing policy is the means to sustain the environment. *Review of Economic Dynamics*, 11(1), 207–219. Elsevier.

Enachi, M. (2009). The knowledge: As production factor. *Studies and Scientific Researches: Economic Edition*. doi: 10.29358/sceco.v0i14.40.

FiBAN. (2018). Finnish startup investments 2018. Finnish Business Angels Network.

Harayama, Y. (2017). Society 5.0: Aiming for a new human-centered society. Collaborative creation through global R&D open innovation for creating the future. *Hitachi Review*, 66(6), 8–13.

Hasan, S., Khan, E. and Nabi, N. (2017). Entrepreneurial education at university level and entrepreneurship development. *Education and Training*, 59. doi: 10.1108/ET-01-2016-0020.

Heal, G. M. et al. (1996). Sustainable use of renewable resources. Available at SSRN: https://ssrn.com/abstract=1613 or http://dx.doi.org/10.2139/ssrn.1613.

Huber, F., Peisl, T., Gedeon, S., Brodie, J. and Sailer, K. (2016). Design thinking-based entrepreneurship education: How to incorporate design thinking principles into an entrepreneurship course. *ECSB Entrepreneurship Education Conference*, Leeds, UK.

Koopmans, T. (1963), *On the Concept of Optimal Economic Growth*, No 163, Cowles Foundation Discussion Papers, Cowles Foundation for Research in Economics, Yale University, https://EconPapers.repec.org/RePEc:cwl:cwldpp:163.

Koski, H. (2011). *Does Marginal Cost Pricing of Public Sector Information Spur Firm Growth?* Helsinki: The Research Institute of the Finnish Economy.

KPMG. (2019). Maharashtra and the exciting growth of its startup ecosystem.

Kurniawan, Y. (2014). The role of knowledge management system in school: Perception of applications and benefits. *Journal of Theoretical and Applied Information Technology*, 61(1), 169–174.

Lucas, R. E. (1988). On the mechanics of economic development. *Journal of Monetary Economics*, 22, 3–42.

Luka, I. (2014). Design thinking in pedagogy. *Journal of Education Culture and Society*, 2, 63–74. doi: 10.15503/jecs20142.63.74.

Mathias, P. (2007). Economic growth and Robinson Crusoe. *European Review*, 15(1), 17–31. doi: 10.1017/S1062798707000038.

Mittal, I. and Gupta, R. K. (2015). Natural resources depletion and economic growth in present era. *SOCH- Mastnath Journal of Science and Technology (BMU, Rohtak)*, 10(3), 24–28.

Moody's Analytics. (2015). Bureau of Labor Statistics. Analysis: Collaborative Economics.

OECD. (2010). Pursuing strong, sustainable and balanced growth: The role of structural reform.

OECD. (2018). Private equity investment in artificial intelligence, OECD Going Digital Policy Note, OECD, Paris. www.oecd.org/going-digital/ai/private-equity-investment-in-artificial-intelligence.pdf.

Phelps, E. (1961). The golden rule of accumulation: A fable for growthmen. *The American Economic Review*, 51(4), 638–643. Retrieved from www.jstor.org/stable/1812790.

Ramsey F. P. (1928). A mathematical theory of saving. *Economic Journal*, 38, 543–559.

Rebelo, S. (1990). Long run policy analysis and long run growth. *Journal of Political Economy*, 99(3), 500–521.

Richardson, K. and Dyball, R. (2016). Needed: new knowledge for sustainable development. *The Solutions Journal*, 7(3), 1–3.

Romer, P. (1990). Endogenous technological change. *Journal of Political Economy*, 98(5), Part 2, S71–S102.

Roncaglia, A. (2005). *The Wealth of Ideas: A History of Economic Thought.* Cambridge, New York: Cambridge University Press.
Saini, T. (1974). Paul Douglas and the Cobb-Douglas production function. *Eastern Economic Journal,* 1(1), 52–58. Retrieved from www.jstor.org/stable/40324601.
Salamzadeh, A. and Kawamorita, H. (2017). The enterprising communities and startup ecosystem in Iran. *Journal of Enterprising Communities People and Places in the Global Economy,* 11.doi: 10.1108/JEC-07-2015-0036.
Solow, R. (1956). A contribution to the theory of economic growth. *The Quarterly Journal of Economic,* 70, 65–94.
Solow, R. (1974). Intergenerational equity and exhaustible resources. *The Review of Economic Studies,* 41, 29–45. Retrieved from www.jstor.org/stable/2296370.
Steigertahl, L. and Mauer, R. (2018). EU startup monitor. 2018 Report. ESCP Europe Jean-Baptiste Say Institute for Entrepreneurship European Commission.
Swan, T. W. (1956). Economic growth and capital accumulation. *Economic Record,* 32, 334–361.
Tahir, L. M. (2013). Creating knowledge practices in school: Exploring teachers knowledge creation. *International Journal of Humanities and Social Science,* 3(1), 147–154.
Uzawa, I. (1965). Optimum technical change in an aggregate model of economic growth. *International Economic Review,* 6, 18–31.
Varian Hal, R. (1992). *Microeconomic Analysis,* Third Edition. New York: W.W. Norton & Company, pp. 431–439.
Vuksanović, D., Vešić, J. and Korčok, D. (2016). Industry 4.0: The future concepts and new visions of factory of the future development. *In Sinteza 2016 - International Scientific Conference on ICT and E-Business Related Research, Belgrade, Singidunum University,* Serbia, pp. 293–298. doi: 10.15308/Sinteza-2016-293-298.
World Economic Forum. (2019). Globalization 4.0 shaping a new global architecture in the age of the fourth industrial revolution. White Paper.

6 Entrepreneurship in Higher Education and Nascent Entrepreneurship

Helen Salavou
Athens University of Economics and Business

CONTENTS

6.1 INTRODUCTION

Entrepreneurship has been and will be a safe vehicle for sustainable development everywhere around the globe. New firms balancing environmental, societal, and economic considerations in the pursuit of an improved quality of life preserve the vitality of national economies. Knowing more about entrepreneurs would be useful. To this end, theory, research, and practice are advanced with individuals being at the center of the entrepreneurial process. Not surprisingly, entrepreneurship education (EE) at universities has gained increasing importance. The development of the EE field emanated from the United States, while later, over the past 20 years, other regions are following with considerable progress, such as Western Europe, Eastern

Europe, Asia, and Latin America. Despite the expansion, the discussion about EE is still in its infancy (Haase and Lautenschläger, 2011) with many questions looking for convincing answers.

This chapter deals with some of the challenging questions in the EE field: (a) If entrepreneurship is teachable, is EE at universities of greater value? (b) Does formal EE make people more skilled? and (c) Can we trace differences among people trying to become entrepreneurs with and without formal EE?

It is true that not everybody has the same entrepreneurial spirit. Another truth is that some people are more entrepreneurial than others. Nevertheless, entrepreneurial skills can be learned, nurtured, and improved. Universities are challenged to educate and qualify entrepreneurs. This chapter focuses on entrepreneurship in higher education and looks for differences between nascent entrepreneurs with and without formal EE. By doing so, it advances our understanding of individuals with and without formal EE in the process of starting a new business. Several scholars have already emphasized the general importance of involving nascent or existing entrepreneurs in the educational process (Carayannis et al., 2003; Collins et al., 2006).

The scope of this chapter is beyond the effects of formal EE on nascent entrepreneurs. Instead, the focal point is to understand who they are. Empirical evidence provided here helps to broaden our perspectives and create new challenges to expand our thoughts.

We emphasize formal EE at universities for at least three reasons. First, graduates are more likely than nongraduates to go on and become successful entrepreneurs (Greene and Saridakis, 2008). Well-educated entrepreneurs are of paramount interest (Raguz and Matic, 2011), as their investment in education is reflected in numerous outcomes (Falkäng and Alberti, 2000; Fayolle et al., 2006; Pittaway et al., 2009; Vesper and Gartner, 1997).

Second, EE has to be considered as an extension of entrepreneurship itself (Sexton and Bowman, 1984), which aims at understanding the phenomenon, learning to become an entrepreneur or an enterprising individual (Farashah, 2013).

Third, EE is seen as a means to stimulate an increased level of economic activity (Farashah, 2013) and speed up entrepreneurship rates, by increasing the formation of new ventures for sustainable development and improving their growth prospects.

The central message from this chapter is that formal EE at the higher level is worth the trouble, at least for some individuals. Three practical tips may benefit individuals flirting with the idea of becoming entrepreneurs and act accordingly:

- *Tip 1: Start the earliest possible:* The earlier you get the knowledge and skills through formal education, the sooner you are in an advantageous position to capitalize on your entrepreneurial efforts.
- *Tip 2: Be aware of what "your people" think about:* Family, as the most common reference group, may be more helpful than you have ever imagined.
- *Tip 3: Go beyond "business as usual," and explore new perspectives:* Take advantage of your special skills, behave politically, and try to do "business for good causes."

This chapter is structured as follows. After this introductory section, the conceptual framework develops thoroughly the arguments leading to three research questions exploring differences in terms of age, family bonds, and political skills between nascent entrepreneurs. In the section of methodology, information about the sample, the data collection, and the statistical analysis is clearly explained. The fourth section discusses the results, whereas the final section provides concluding remarks in the form of practical tips.

6.2   CONCEPTUAL FRAMEWORK

The EE literature embraces scholars dealing with any pedagogical program or process of education for entrepreneurial competencies, which involves developing certain personal qualities (Fayolle et al., 2006). Although EE takes place in a plethora of educational entities across countries based on various criteria (e.g., objectives, target audiences), this chapter focuses on higher education institutions due to the fundamental role they play (Haase and Lautenschläger, 2011). Universities traditionally represent the main source of new knowledge and hold a constantly regenerating stock of students and scientists. They are challenged to endow everyone actively with the appropriate motivation, knowledge, and abilities for firm creation (Haase and Lautenschläger, 2011). Studies have shown that individuals with a university degree have a higher proclivity to start their own business (Sternberg et al., 2007). Furthermore, students after EE are expected to perform at levels above that of being an employee (Klandt, 2004) and show career progression on self-employment.

Since its starting point back in 1947 at Harvard Business School, entrepreneurship in higher education has reported a growing interest in many countries (Raguz and Matic, 2011). Beyond the United States, which is the pace-setting country in the EE field, considerable progress has been made in countries of Western Europe, Eastern Europe, Asia, and Latin America. Despite confusion between the terms used for education in the entrepreneurship field and after many changes based on regional preferences, the recent conceptual work of Haase and Lautenschläger (2011) concludes that the scientific literature has primarily adopted the term "entrepreneurship education" as dominant. However, there is still a fundamental concern over terminology, which creates an abundant heterogeneity regarding philosophy, objectives, content methodologies, and effectiveness. As the development of overarching standards for EE is advisable (Haase and Lautenschläger, 2011), we clarify the positioning of this work before we develop the research questions.

First, among the two approaches answering the "old" question: Is entrepreneurship teachable? (Haase and Lautenschläger, 2011), this work favors the behavioral approach (entrepreneurs can be taught and educated) and not the trait approach (entrepreneurs have a unique personality and a fixed state of existence; traits can neither be learned nor be developed through education, training, or professional experience).

Second, this work deals with the mode of EE embracing "education for enterprise" (preparation for self-employment) and not as Jamieson (1984) reports "education about enterprise" (to create awareness) or "education in enterprise" (management training). We agree with scholars claiming that EE should be designed as close to reality as possible, emulating contexts similar to those in which entrepreneurs act (Carayannis et al., 2003; Hindle, 2004), including methodologies, such as working with entrepreneurs (Hills, 1988; Johannisson, 1991; Carayannis et al., 2003) and developing actual business start-ups (Hills, 1988; Truell et al., 1998).

Third, this work is in line with Haase and Lautenschläger (2011) that EE must primarily target a change in the individual's entrepreneurial "know-how" (soft skills) as opposed to "know-what" (hard facts) and "know-why" (conviction). Universities still have a long way to go, as only a few approaches are suitable to transmit the entrepreneurial "know-how." According to Kirby (2004), the focus on developing entrepreneurial skills, attributes, and behavior is rather scarce. Academics should desist from simply teaching hard facts about business creation and management (know-what). Instead, they should get closer to the notions of entrepreneurial doing, thinking, and feeling (know-why), which will open space for experiencing entrepreneurship as the only way to convey the necessary "know-how."

The discussion below responds to the challenge of knowing more about individuals in the process to establish a new business along with the role of entrepreneurship in higher education.

6.2.1 EDUCATION IN ENTREPRENEURSHIP AND AGE

The importance of EE has increased due to the need to prepare students for coping in the contemporary work and living environment (Küttima et al., 2014). Empirical studies emphasizing entrepreneurship as an employment choice not only include factors such as education and age to be important drivers of entrepreneurial behavior but also produce evidence from samples of individuals in the process of starting a new venture (Arenius and Minniti, 2005). Not surprisingly, the relationship between age and the likelihood of starting a new business picks at a relatively early age and decreases thereafter (Lévesque and Minniti, 2006). More specifically, empirical evidence (Reynolds et al., 2003) shows that individuals are most likely to be nascent entrepreneurs between 25 and 34 years of age. Similarly, the likelihood of being a nascent entrepreneur is maximized among young individuals, whereas the probability of being an entrepreneur is highest among older individuals (Blanchflower, 2004). All in all, entrepreneurship is not only, as Arenius and Minniti (2005) report, a young's man game but also a game played during the period of adult education. Wilson et al. (2007) argue that EE can also increase students' interest in entrepreneurship as a career.

In the absence of empirical evidence on nascent entrepreneurs of different ages, especially when the educational background is involved, a relevant research question follows.

Research Question 1:
Are nascent entrepreneurs with and without formal EE of different ages?

6.2.2 EDUCATION IN ENTREPRENEURSHIP AND FAMILY BONDS

The entrepreneurship literature deals with family as the most common reference group that shapes individual beliefs (Ajzen, 1991). Why? This is because the family has a dominant presence in the social environment of every individual. The social pressure that an individual expects to receive based on how well he/she performs has been the subject of extensive research (Ajzen, 1991). Many subfields in the entrepreneurship literature are quite often concerned with the role of the family in determining the entrepreneurial career choice of individuals (Dyer, 1995). According to Gartner (1985), entrepreneurs differ from non-entrepreneurs in social characteristics, such as family. It appears, for example, that family support (i.e., positive perceptions about an entrepreneurial career) influences individuals' perceptions of their ability to become entrepreneurs (Bandura, 1997; Carsrud and Brännback, 2011). In addition, the desire to be an entrepreneur is influenced by sociocultural elements, such as family (Summers, 1998).

The role of education in affecting subjective norms (including perceived family expectations and beliefs) merits further investigation (Basu and Virick, 2008). Souitaris et al. (2007) found that entrepreneurship programs significantly raise students' subjective norms by inspiring them to choose entrepreneurial careers. On the contrary, Basu and Virick (2008) did not provide empirical support that students with prior exposure to EE have more positive attitudes toward entrepreneurship or that stronger family support favors entrepreneurship.

Responding to the need for more empirical studies focusing on higher education (e.g., Nabi et al., 2017), a relevant research question follows.

Research Question 2:
    Are nascent entrepreneurs with and without formal EE different in terms of family bonds?

6.2.3 EDUCATION IN ENTREPRENEURSHIP AND POLITICAL SKILLS

Skills development is continuous (Chell, 2013). Universities are challenged to convey a broad range of skills (i.e., social, technical, conceptional) on how to start and run a business, considering that they suggest key elements for success (Haase and Lautenschläger, 2011). Carter et al. (2003) argue that differences in educational level and skills could explain why some people have successful entrepreneurial behavior, while others do not.

Some skills are determined by intelligence, while others, like political skills, are independent from general mental ability. Politically skilled individuals convey a sense of personal security and calm self-confidence that attracts others and gives them a feeling of comfort (Ferris et al., 2005). Because their focus is outward toward others, individuals maintain a proper balance between their accountability to others and themselves. It is, however, meaningless to talk of skills in a vacuum, without reference to a context (Spenner, 1990). Political skills do not address the "whole person" but aspects where engagement with others is important (Chell, 2013). They appear to be directly relevant to activities performed by entrepreneurs, namely, activities that play an important role in the survival and success of their new ventures (Baron and Tang, 2009).

Political skills are among those entrepreneurial skills enabling nascent entrepreneurs to be more effective in business start-up (Bird, 1989; Markman, 2007).

Scholars raise assumptions that political skills are open to modification through training or other intervention (Baron and Tang, 2009) or can partially be developed or shaped through formal or informal experiences (Ferris et al., 2002). In the absence of empirical evidence along with recent calls for research on the political skill of the influencer, which is ill-defined, a relevant research question follows.

Research Question 3:
Are nascent entrepreneurs with and without formal EE different in terms of political skills?

6.3 METHODOLOGY

6.3.1 SAMPLE AND DATA COLLECTION

The sample of this empirical study consists of 203 randomly selected Greek adults with an interest in entrepreneurship as a professional choice. More specifically, adults are classified into two groups, with and without formal EE in higher education. The former group consists of 62 adults having attended entrepreneurship courses during their postgraduate studies. The latter group comprises 141 adults without formal EE at universities, who are participants in start-up contests, business incubators, and/or accelerators. In the context of a Greek research study (Salavou, 2017), data were collected through a structured online questionnaire sent by email to the respondents along with instructions.

Table 6.1 provides the sample characteristics. From the respondents, 61% are males and 39% are females. The average age of the respondents is 34 years. Furthermore, 58% of them have completed postgraduate studies (31% have attended entrepreneurship courses), and 10% hold a PhD degree. Only a small percentage of their parents have completed postgraduate and doctoral studies, while 55% of their mothers and 62% of their fathers are still active in the marketplace. Almost one-third of their parents follow entrepreneurship as a professional career, which is more than what is typical for Greece, namely, one-fifth (Endeavor, 2013).

TABLE 6.1
Sample Characteristics

No of persons	203
Males (%)	61.08
Age	34.00
Educational Level	
Postgraduate studies (%)	57.64
PhD	9.85
High volunteers (%)	73.40
Goal to establish new venture (%)	88.18
Goal to establish new social enterprise (%)	94.58

(*Continued*)

TABLE 6.1 (*Continued*)
Sample Characteristics

Parents' Educational Level

Mother

Postgraduate studies (%)	6.90
PhD	1.48

Father

Postgraduate studies (%)	9.36
PhD	2.96

Parents' Professional Status

Mother

Working (%)	55.17
Retired (%)	35.47
Unemployed (%)	9.36

Father

Working (%)	62.18
Retired (%)	36.82
Unemployed (%)	1.00
One of the parents' entrepreneur	36.45

6.3.2 STATISTICAL ANALYSIS

STATA 13.0 software was used to conduct the statistical analysis. The Appendix provides the measures used in this study and the factor analysis to verify their validity. Following the confirmation of the construct validity, averages of items pertaining to factors extracted are used to form the variables for further statistical analysis. Table 6.2 presents descriptive statistics together with the inter-item reliability coefficient of the variables, which are acceptable according to the organizational attribute reliability standards suggested by Van de Ven and Ferry (1980).

To test the three research questions of the conceptual framework, we use a one-way analysis of variance (ANOVA). More specifically, we use the two groups of nascent entrepreneurs, with and without formal EE, as independent variables and the variables under investigation as dependent variables. Tables 6.3, 6.6, and 6.8 report the results to confirm or reject the research questions. In a rather supportive way, we use supplementary analysis to better explain the results referring to age, family bonds, and political skills. Consequently, additional tables are provided in case of statistically significant results. These tables refer to age (see Tables 6.4 and 6.5), family bonds (see Table 6.7), and political skills (see Tables 6.9–6.12) and report the results from using filters to further split the two subsamples of adults in terms of venture creation focus (1: low focus; 2: high focus), volunteerism tendency (1: low; 2: high), and gender (men versus women). We wanted to further explore whether adults

TABLE 6.2

Descriptives and Reliability Coefficients

Variable	Dimensions	Composite Scores	Mean	SD	Min	Max	Alpha[a]
Age (in years)		Continuous variable	34	8	21	65	na
Family bonds		Average of items extracted from factor analysis	5.29	1.17	2	7	0.73
Political skills		Average of items extracted from factor analysis					
	Social astuteness		5.47	1.30	1	7	0.79
	Humanistic astuteness		4.99	1.69	1	7	0.73
	Ecological astuteness		5.64	1.14	2	7	0.65
	Economic astuteness		3.09	1.71	1	7	na

[a] na = non-applicable.

(a) prioritizing to become entrepreneurs have differences from others, (b) emphasizing volunteerism have differences from others, and (b) are different based on gender. The discussion comments on statistically significant differences ($p < 0.01$, $p < 0.05$, and $p < 0.10$) without using statistical terminology.

6.4   RESULTS AND DISCUSSION

This section provides answers to three research questions, which advance the understanding of nascent entrepreneurs with formal EE. Regarding the first research question exploring differences in terms of age, the findings indicate (see Table 6.3) that nascent entrepreneurs with formal EE are younger than those without formal EE. More specifically, individuals at the age of 31 have already completed postgraduate studies to cultivate their entrepreneurial skills.

Additional evidence from supplementary analysis reveals that nascent entrepreneurs with formal EE, who prioritize entrepreneurship as a profession (see Tables 6.4 and 6.5), are of younger age (31 years old).

To summarize, our findings corroborate the work of Reynolds et al. (2003) that age plays a role for young adults experiencing the entrepreneurial process as nascent entrepreneurs. This is in line with dominant evidence that age differentiates

TABLE 6.3

Differences between Nascent Entrepreneurs with and without Formal EE

	with Formal EE[a]	without Formal EE[a]	F	p-Value[b]
Age (in years)	30.82	34.35	6.75	0.01

[a] Figures represent the mean values in each cluster.
[b] Significance level (p-value) is based on one-way ANOVA.

TABLE 6.4
Differences between Nascent Entrepreneurs Based on Venture Creation Focus

	with Formal EE[a]	without Formal EE[a]	F	p-Value[b]
Age (in years)	30.82	34.35	6.75	0.01

[a] Figures represent the mean values in each cluster.
[b] Significance level (p-value) is based on one-way ANOVA.

TABLE 6.5
Differences between Nascent Entrepreneurs with Formal EE

	Low Focus on Venture Creation[a]	High Focus on Venture Creation[a]	F	p-Value[b]
Age (in years)	35.85	30.82	4.89	0.03

[a] Figures represent the mean values in each cluster.
[b] Significance level (p-value) is based on one-way ANOVA.

entrepreneurs from the general population (Begley and Boyd, 1987; Brockhaus, 1980, 1982; Brockhaus and Horwitz, 1986; Sexton and Bowman, 1984). Our findings show that the younger the nascent entrepreneurs, the more they are looking for formal education to advance their entrepreneurial capacity. In addition, younger ages prefer entrepreneurship as a profession. Overall, the evidence reported here supports assumptions (Stuetzer et al., 2013) that those who develop a first entrepreneurial career interest earlier in life may start earlier to invest in a balanced skill set to be more skilled as nascent entrepreneurs.

Regarding the second research question exploring differences in terms of family bonds, Table 6.6 reveals that nascent entrepreneurs with formal EE have slightly a need for more family support than those without formal EE. It appears that for some young adults, the family is indeed a reference group that shapes individual beliefs (Ajzen, 1991) or, stated differently, the family plays indeed a role for choices relevant to the entrepreneurial career of individuals (Dyer, 1995; Souitaris et al., 2007).

TABLE 6.6
Differences between Nascent Entrepreneurs with and without Formal EE

	with Formal EE[a]	without Formal EE[a]	F	p-Value[b]
Family bonds	5.49	5.20	2.84	0.09

[a] Figures represent the mean values in each cluster.
[b] Significance level (p-value) is based on one-way ANOVA.

Additional evidence from supplementary analysis indicates that nascent entrepreneurs without formal EE, who prioritize entrepreneurship as a profession (see Table 6.7), are also in need of stronger family support. Consequently, for less-skilled individuals, who really want to be entrepreneurs, family members act as a strong driving force.

Regarding the third research question exploring differences in terms of political skills, we did not find any (Table 6.8). At first glance, all dimensions of political skills, namely, social, humanistic, ecological, and economic astuteness, are of equal importance for all nascent entrepreneurs (with and without formal EE).

Additional evidence from a supplementary analysis looking deeper into the two groups of nascent entrepreneurs is noteworthy. First, differences between them appear if new venture creation is a priority (see Tables 6.9 and 6.10). More specifically, nascent entrepreneurs with formal EE (see Table 6.9) have higher social astuteness, such as making errands for friends in need and offering help for people in need. On the contrary, nascent entrepreneurs without formal EE (see Table 6.10) have higher ecological astuteness, such as showing concern for problems on the earth, recycling waste, and using carefully electricity.

Second, further gender-specific evidence shows differences only among nascent entrepreneurs without formal EE in terms of their humanistic astuteness (see Table 6.11). More specifically, women are more sensitive in human aspects than men. This

TABLE 6.7
Differences between Nascent Entrepreneurs without Formal EE

	Low Focus on Venture Creation[a]	High Focus on Venture Creation[a]	F	p-Value[b]
Family bonds	4.24	5.28	7.21	0.00

[a] Figures represent the mean values in each cluster.
[b] Significance level (p-value) is based on one-way ANOVA.

TABLE 6.8
Differences between Adults with and without Formal EE

	Adults with EE[a]	Adults without EE[a]	F	p-Value[b]
	Political Skills			
Social astuteness	5.35	5.52	0.78	ns
Humanistic astuteness	5.00	4.99	0.00	ns
Ecological astuteness	5.67	5.62	0.09	ns
Economic astuteness	3.26	3.01	0.88	ns

[a] Figures represent the mean values in each cluster.
[b] Significance level (p-value) is based on one-way ANOVA, ns = nonsignificant.

TABLE 6.9
Differences between Nascent Entrepreneurs with Formal EE

	Low Focus on Venture Creation[a]	High Focus on Venture Creation[a]	F	p-Value[b]
	Political Skills			
Social astuteness	4.31	5.63	10.32	0.00

[a] Figures represent the mean values in each cluster.
[b] Significance level (p-value) is based on one-way ANOVA.

TABLE 6.10
Differences between Nascent Entrepreneurs without Formal EE

	Low Focus on Venture Creation[a]	High Focus on Venture Creations[a]	F	p-Value[b]
	Political Skills			
Ecological astuteness	4.91	5.68	4.78	0.03

[a] Figures represent the mean values in each cluster.
[b] Significance level (p-value) is based on one-way ANOVA.

TABLE 6.11
Differences between Nascent Entrepreneurs without Formal EE

	Men[a]	Women[a]	F	p-Value[b]
Humanistic astuteness	4.71	5.43	9.08	0.00

[a] Figures represent the mean values in each cluster.
[b] Significance level (p-value) is based on one-way ANOVA.

finding might further explain why men are twice as likely as women to become entrepreneurs (Allen et al., 2008; Wilson et al., 2004; Zellweger et al., 2010). However, additional research would help to verify whether EE is gender sensitive (Wilson et al., 2007).

Third, the final evidence comes from differences among nascent entrepreneurs with formal EE if volunteerism tendency is involved (see Table 6.12). More specifically, those inclined to volunteerism have higher social astuteness (such as making errands for friends in need and offering help for people in need) and slightly higher humanistic astuteness (such as donating money and being aware of disasters) and ecological astuteness (such as showing concern for problems on the earth, recycling waste, and using carefully electricity).

TABLE 6.12
Differences between Nascent Entrepreneurs with Formal EE

	Low Volunteerism Tendency[a]	High Volunteerism Tendency[a]	F	p-Value[b]
	Political Skills			
Social astuteness	4.62	5.78	11.30	0.00
Humanistic astuteness	4.54	5.27	3.74	0.06
Ecological astuteness	5.34	5.87	3.23	0.08

[a] Figures represent the mean values in each cluster.
[b] Significance level (p-value) is based on one-way ANOVA.

To summarize the empirical findings in terms of political skills, it appears that:

• Nascent entrepreneurs with and without formal EE have differences in some political skills;
• Among nascent entrepreneurs with formal EE, volunteers behave more politically in social, humanistic, and ecological aspects;
• Nascent entrepreneurs, regardless of differing education, have the same economic astuteness;
• Gender plays a role only for nascent entrepreneurs without EE, when behaving politically in terms of human aspects.

Overall, this work reflects the richness of empirical findings while exploring the political skills of entrepreneurs. More studies are needed to explain why formal EE qualifies nascent entrepreneurs with some political skills and/or why nascent entrepreneurs with certain political skills prefer formal, as opposed to informal EE. The question whether the full range of political skills is teachable deserves further empirical investigation. In accordance with Haase and Lautenschläger (2011), conveying entrepreneurial soft skills, such as political, is a difficult undertaking, but highly relevant for nascent and actual entrepreneurs (Haase and Lautenschläger, 2011).

6.5 CONCLUDING COMMENTS

Are entrepreneurs made or born? Whatever the answer is, individuals can learn, nurture, and improve entrepreneurial skills. This is an indisputable truth mirrored in myriad activities, projects, and strategic policies to educate and qualify potential and actual entrepreneurs at the national and international level. This chapter responds to the need for a deeper understanding of young people in the process to start a new venture. By investigating two groups, one with and one without formal education in entrepreneurship, we provide empirical evidence on differences in terms of age, family bonds, and political skills.

The central message from this chapter is that formal education in entrepreneurship at a higher level is worth the trouble. However, it is not an option for everyone. The "one-size-fits-all" EE approach may not be appropriate (Wilson et al., 2007). Formal EE is most attractive for those who are younger, sustain stronger family bonds, and are skilled at behaving politically. Three meaningful implications follow in the form of "practical tips." They serve the need to broaden our perspectives and create new challenges to expand our thoughts.

6.5.1   TIP 1: START AS EARLY AS POSSIBLE

The question "when should I start my own venture?" can receive millions of answers. Nevertheless, the earlier you get the knowledge and skills through formal education, the sooner you are in an advantageous position to capitalize on your entrepreneurial efforts. Why? First, individuals who develop an entrepreneurial career interest as early as in adolescence more often engage in entrepreneurship during their subsequent career than others (Falck et al., 2012; Schmitt-Rodermund, 2004). Second, a significant percentage of university graduates forms high-quality start-ups shortly after graduation (Åstebro et al., 2012; Holden and Jameson, 2002). Third, the younger individuals are most likely to engage in activities necessary to start a new venture (Kelly et al., 2010).

6.5.2   TIP 2: BE AWARE OF WHAT "YOUR PEOPLE" THINK ABOUT

The question "Do I tell my family I want to be an entrepreneur?" can have both a positive or negative answer. However, if you are forced to give only one answer, prefer to say "yes." Why? First, family is the most common reference group for every individual beliefs (Ajzen, 1991). Second, family (Dyer, 1995) as well as role models are important in the decision to start a business (Krueger and Carsrud, 1993). Why looking for advice outside your family if you already have some? Whether or not you feel like, family members (even close friends) may be more helpful than you have ever imagined. Don't lose time for nothing. They can play an advisory role along a promising professional future.

6.5.3   TIP 3: GO BEYOND "BUSINESS AS USUAL"
AND EXPLORE NEW PERSPECTIVES

Take advantage of your special skills and explore new perspectives while enterprising. Do not be afraid to behave politically. Members of society all over the continent call for considerate responses and creative endeavors addressing social and environmental concerns at an unparalleled scale. Achieving "sustainable development" requires balancing environmental, societal, and economic considerations in the pursuit of an improved quality of life. Think, for example, about the triple-bottom-line approach (Elkington, 1997). It is a rather up-to-dated approach, which helps organizations to balance economic propensity, environmental integrity, and social justice (McDonald et al., 2015). In other words, it does not "sacrifice" tangible financial returns to maximize the creation of less-tangible social and/or environmental value (Haigh and Hoffman, 2012). Why not grasping such a challenge if you are skilled enough to do it? Do not waste your time. Go beyond "business as usual," and try to do "business for good causes." It might be rewarding.

APPENDIX

TABLE A.1
Measurement Scales

Variables	Items	References
Family bonds	*Please indicate (from 1 to 7) your level of agreement with the following statements:* 1. There are strong ties with family members. 2. I take seriously family advice. 3. My family's social networking helps me to achieve my goals.	Created by the author based on the work of Kolvereid (1996)
Political skills	1. I have made errands for my friends when they got sick. 2. I offer my help to people I hardly know. 3. I help people in need. 4. I am a volunteer. 5. I donate money for natural disasters (i.e., earthquake). 6. I like to know when we experience crises or disasters. 7. Petroleum leaks and chemical waste cause serious problems on the earth. 8. Environmental issues follow economic growth. 9. I always recycle waste at home. 10. In winter, I never set the heating to its maximum setting. 11. When I brush my teeth, I turn off the tap.	Created by the author based on the work of Ferris et al. (2005)
Gender	1: male; 2: female	
Age	Continuous variable in years	
Volunteer	1: low; 2: high	
Goal (to become entrepreneur)	1: low; 2: high	

TABLE A.2
Family Bonds Scale Factor Analysis

Items	Factor Loadings
There are strong ties with family members	0.80
I take seriously family advice	0.90
My family's social networking helps me to achieve my goals	0.71
Proportion of variance	**0.65**

TABLE A.3
Political Skills Scale Factor Analysis

Items	Factor Loadings
Social Astuteness	
I have made errands for my friends when they got sick	0.74
I offer my help to people I hardly know	0.89
I help people in need	0.81
Humanistic Astuteness	
I donate money for natural disasters (i.e., earthquake)	0.88
I like to know when we experience crisis or disasters	0.84
Ecological Astuteness	
Petroleum leaks and chemical waste cause serious problems on the earth	0.62
I always recycle waste at home	0.63
In winter, I never set the heating to its maximum setting	0.78
When I brush my teeth, I turn off the tap	0.65
Economic Astuteness	
Environmental issues follow economic growth	0.97
Proportion of Variance	**0.68**

REFERENCES

Ajzen, I. (1991). The theory of planned behavior. *Organizational Behavior and Human Decision Processes*, 50 (2): 179–211.

Allen, I.E., Elam, A., Langowitz, N. and Dean, M. (2008). Global entrepreneurship monitor. 2007 Report on Women and Entrepreneurship: The Center for Women's Leadership at Babson College, Babson Park, MA.

Arenius, P. and Minniti. M. (2005). Perceptual variables and nascent entrepreneurship. *Small Business Economics*, 24: 233–247.

Åstebro, T., Bazzazian, N. and Braguinsky, S. (2012). Startups by recent university graduates and their faculty: Implications for university entrepreneurship policy. *Research Policy*, 41: 663–677.

Bandura, A. (1997). *Self-Efficacy: The Exercise of Control*. New York: Freeman.

Baron, R.A. and Tang, J. (2009). Entrepreneurs' social skills and new venture performance: Mediating mechanisms and cultural generality. *Journal of Management*, 35 (2): 282–306.

Basu, A. and Virick, M. (2008). Assessing entrepreneurial intentions amongst students: A comparative study. Available online on http://nciia.org/conf08/assets/pub/basu2.pdf.

Begley, T.M. and Boyd, D.P. (1987). Psychological characteristics associated with performance in entrepreneurial firms and smaller businesses. *Journal of Business Venturing*, 2(1): 79–93.

Bird, B.J. (1989). *Entrepreneurial Behavior*. Glenview, IL: Scott, Foresman & Co.

Blanchflower, D.G. (2004). Self-employment: More may not be better. NBER Working Paper No. 10286.

Brockhaus, R.H. (1980). Risk taking propensity of entrepreneurs. pp. 509–520.

Brockhaus, R.H. (1982). The psychology of the entrepreneur. In: *Encyclopedia of Entrepreneurship* (pp. 39–57).

Brockhaus, R.H. and Horwitz, P.S. (1986). The art and science of entrepreneurship. In: Sexton, X.L. and Smilor, R.W. *The Psychology of the Entrepreneur* (pp. 25–48). Cambridge, MA: Ballinger.

Carayannis, E.G., Evans, D. and Hanson, M. (2003). A cross-cultural learning strategy for entrepreneurship education: Outline of key concepts and lessons learned from a comparative study of entrepreneurship students in France and the US. *Technovation*, 23(9): 757–771. doi: 10.1016/S0166-4972(02)00030-5.

Carsrud, A. and Brännback, M. (2011). Entrepreneurial motivations: What do we still need to know? *Journal of Small Business Management*, 49(1): 9–26.

Carter, N.M., Gartner, W.B., Shaver, K.G. and Gatewood, E.J. (2003). The career reasons of nascent entrepreneurs. *Journal of Business Venturing*, 18(1): 13–39, doi: 10.1016/S0883-9026(02)00078-2.

Chell, E. (2013). Review of skill and the entrepreneurial process. *International Journal of Entrepreneurial Behavior and Research*, 19 (1): 6–31. doi: 10.1108/13552551311299233.

Collins, L.A., Smith, A.J. and Hannon, P.D. (2006). Applying a synergistic learning approach in entrepreneurship education. *Management Learning*, 37(3): 335–354. doi: 10.1177/1350507606067171.

Dyer, W.G. (1995). Toward a theory of entrepreneurial careers. *Entrepreneurship Theory and Practice*, 19 (2): 7–21.

Elkington, J. (1997). *Cannibals with Forks: The Tripple Bottom Line of 21st Century Business*. Reading: Capstone Publishing Ltd.

Endeavor (2013). Endeavor greece. www.endeavor.org.gr Greece.

Falck, O., Heblich, S. and Luedemann, E. (2012). Identity and entrepreneurship: Do school peers shape entrepreneurial intentions? *Small Business Economics*, 39 (1): 39–59. doi: 10.1007/s11187-010-9292-5.

Falkäng, J. and Alberti, F. (2000). The assessment of entrepreneurship education. *Industry and Higher Education*, 14 (2): 101–108.

Farashah, D.A. (2013). The process of impact of entrepreneurship education and training on entrepreneurship perception and intention. *Education + Training*, 55 (8/9): 868–885.

Fayolle, A., Gailly, B. and Lassas-Clerc, N. (2006). Assessing the impact of entrepreneurship education programmes: A new methodology. *Journal of European Industrial Training*, 30 (9): 701–720.

Ferris, G.R., Hochwarter, W.A., Douglas, C., Blass, F.R., Kolodinsky, R.W., and Treadway, D.C. (2002). Social influence processes in organizations and human resources systems. In: Ferris, G.R. and Martocchio, J.J. (Eds.) *Research in Personnel and Human Resources Management* (vol. 21, pp. 65–127). Oxford: JAI/Elsevier Science.

Ferris, G.R., Treadway, D.C., Kolodinsky, R.W., Hochwarter, W.A., Kacmar, C.J., Douglas, C. and Frink, D.D. (2005). Development and validation of the political skill inventory. *Journal of Management*, 31 (1): 126–152.

Gartner, W.B. (1985). A conceptual framework for describing the phenomenon of new venture creation. *Academy of Management Review*, 10 (4): 696–706.

Greene, F.J. and Saridakis, G. (2008). The role of higher education skills and support in graduate self-employment. *Studies in Higher Education*, 33 (6): 653–672.

Haase, H. and Lautenschläger, A. (2011). The 'Teachability Dilemma' of entrepreneurship. *International Entrepreneurship Management Journal*, 7: 145–162. doi: 10.1007/s11365-010-0150-3.

Haigh, N. and Hoffman, A.J. (2012). Hybrid organizations: The next chapter of sustainable business. *Organizational Dynamics*, 41, 126–134.

Hills, G.E. (1988). Variations in University entrepreneurship education: An empirical study of an evolving field. *Journal of Business Venturing*, 3(2): 109–122.

Hindle, K. (2004). A practical strategy for discovering, evaluating, and exploiting entrepreneurial opportunities: Research-based action guidelines. *Journal of Small Business and Entrepreneurship*, 17 (4): 267–276.

Holden, R. and S. Jameson. (2002). Employing graduates in SMEs: Towards a research agenda. *Journal of Small Business and Enterprise Development*, 9: 271–284.

Jamieson, I. (1984). Education for enterprise. In: Watts, A.G. and Moran, P. (Eds.) *CRAC* (pp. 19–27). Cambridge, MA: Ballinger.

Johannisson, B. (1991). University training for entrepreneurship: Swedish approaches. *Entrepreneurship and Regional Development*, 3(1): 67–82. doi: 10.1080/08985629100000005.

Kelly, D.J., Bosma, N. and Amorós, J.E. (2010). Global entrepreneurship monitor: 2010 global report. Available at www.gemconsortium.org/docs/266/gem-2010-global-report.

Kirby, D.A. (2004). Entrepreneurship education can business schools meet the challenge. *Education + Training*, 46(8): 510–519.

Klandt, H. (2004). Entrepreneurship education and research in German-speaking Europe. *Academy of Management Learning and Education*, 3 (3): 293–301.

Kolvereid, L. (1996). Organisational employment versus self employment: Reasons for career choice intentions. *Entrepreneurship Theory and Practice*, 20(3): 23–31.

Krueger, N. F. and Carsrud, A. L. (1993). Entrepreneurial intentions: Applying the theory of planned behaviour. *Entrepreneurship and Regional Development*, 5(4): 315–330.

Küttima, M., Kallastea, M., Venesaara, U. and Kiisb, A. (2014). Entrepreneurship education at university level and students' entrepreneurial intentions. *Procedia: Social and Behavioral Sciences*, 110: 658–668. doi: 10.1016/j.sbspro.2013.12.910.

Lévesque, M. and Minniti, M. (2006). The effect of aging on entrepreneurial behavior. *Journal of Business Venturing*, 21(2): 177–194.

Markman, G.D. (2007). Entrepreneurs' competencies. In: Robert, B., Michael, F. and Robert, B. (Eds) *The Psychology of Entrepreneurship* (pp. 67–92). Mahwah, NJ and London: Robert Erlbaum.

McDonald, R., Weerawardena, J., Madhavaram, S. and Sullivan Mort, G. (2015). From "virtuous" to "pragmatic" pursuit of social mission. *Management Research Review*, 38 (9): 970–991. doi: 10.1108/MRR-11-2013-0262.

Nabi, G., Liñán, F., Fayolle, A., Krueger, N. and Walmsley, A. (2017). The impact of entrepreneurship education in higher education: A systematic review and research agenda. *Academy of Management Learning and Education*, 16 (2): 277–299.

Pittaway, L., Hannon, P., Gibb, A. and Thompson, J. (2009). Assessment practice in enterprise education. *International Journal of Entrepreneurial Behavior and Research*, 15 (1): 71–93.

Raguz, I.V. and Matic, M. (2011). Student's perceptions and intentions towards entrepreneurship: The empirical findings from the University of Dubrovnik - Croatia. *International Journal of Management Cases*, 13 (3): 38–49.

Reynolds, P.D., Bygrave, B. and Hay, M. (2003). *Global Entrepreneurship Monitor Report*, Kansas City, MO: E. M. Kauffmann Foundation.

Salavou, H. (2017). Determinants of Young Entrepreneurship in Greece during Financial Crisis, Project of Original Scientific Publications of Athens University of Economics and Business Faculty Members (Participation as scientific director), Research Centre of the Athens University of Economics and Business, funded by national resources, Greece.

Schmitt-Rodermund, E. (2004). Pathways to successful entrepreneurship: Parenting, personality, competence, and interests. *Journal of Vocational Behavior*, 65 (3): 498–518.

Sexton, D.L. and Bowman, N.B. (1984). Entrepreneurship education: Suggestions for increasing effectiveness. *Journal of Small Business Management*, 22 (000002): 18.

Souitaris, V., Zerbinati, S. and Al-Laham, A. (2007), Do entrepreneurship programmes raise entrepreneurial intention of science and engineering students? The effect of learning, inspiration and resources. *Journal of Business Venturing*, 22(4): 566–591.

Spenner, K.I. (1990). Skill: Meanings, methods, and measures. *Work and Occupations*, 17 (4): 399–421.

Sternberg, R., Brixy, U. and Hundt, C. (2007). *Global Entrepreneurship Monitor. Länderbericht Deutschland 2006.* Hannover/Nürnberg: Global Entrepreneurship Research Association.

Stuetzer, M., Obshonka, M. and Schmitt-Rodermund, E. (2013). Balanced skills among nascent entrepreneurs. *Small Business Economics*, 41: 93–114. doi: 10.1007/s11187-012-9423-2.

Summers, D.F. (1998). An empirical investigation of personal and situational factors that relate to the formation of entrepreneurial intentions. Unpublished doctoral dissertation. University of North Texas.

Truell, A.D., Webster, L. and Davidson, C. (1998). Fostering the entrepreneurial spirit: Integrating the business community into the classroom. *Business Education Forum*, 53(2): 28–29.

Van de Ven, A. and Ferry, D. (1980). *Measuring and Assessing Organizations.* New York: Wiley.

Vesper, K.H. and Gartner, W.B. (1997). Measuring progress in entrepreneurship education. *Journal of Business Venturing*, 12 (5): 403–421.

Wilson, F., Kickul, J. and Marlino, D. (2007). Gender, entrepreneurial self-efficacy, and entrepreneurial career intentions: Implications of entrepreneurship education. *Entrepreneurship: Theory and Practice*, 31 (3): 387–406.

Wilson, F., Marlino, D. and Kickul, J. (2004). Our entrepreneurial future: Examining the diverse attitudes and motivations of teens across gender and ethnic identity. *Journal of Developmental Entrepreneurship*, 9(3): 177.

Zellweger, T., Sieger, P. and Halter, F. (2010). Should I stay or should I go? Career choice intentions of students with family business background. *Journal of Business Venturing*, 26(5): 521–536.

7 Education in Social Entrepreneurship for Sustainable Development

A Case Study in Teacher Education at the University of Athens, Greece

Yiannis Roussakis
University of Thessaly

Vasiliki Brinia
Athens University of Economics and Business

Thalia Dragonas
National and Kapodistrian University of Athens

CONTENTS

7.1   INTRODUCTION: ENTREPRENEURSHIP EDUCATION IN CONTEMPORARY KNOWLEDGE SOCIETY

Since the 1990s, "entrepreneurship education" has been, increasingly, among the most discussed subjects of the educational agenda worldwide. It has become a universalized "keyword" (Vavrus (2004: 142), an integral part of what Drori, Jang, and Meyer (2006: 220) call a globalized "shared script," which prescribes the values, content, organizational forms, and expected outcomes of education, in national and international settings. Following this narrative, education is expected, among others, to move beyond traditional practices, and enable students to "map out the new pathways that correspond to current realities" (Auerswald, 2012: 186), as prospects of employment and well-being are increasingly related to the ability of individuals to continue to learn and develop their competences (EC, 2007); to prioritize creativity and innovation (Wagner and Compton, 2012); to produce new forms of individual and social "responsibility" (Bauman, 2005: 115) within the contemporary "risk society" (Beck, 1992); to nurture new identities of entrepreneurial citizens and citizens-consumers (Rose and Osborne, 2000); to raise awareness for global and local problems and become part of their solution; and to facilitate the emergence of an "entrepreneurial self" and of a transformed concept of *homo economicus* (Besley and Peters, 2007: 155).

In view of the above, growing attention for "entrepreneurship education" relates to a wide array of factors impacting its form, content, and focus. One set of factors pertains to the rise of neoliberal globalization, the emergence of knowledge economy, and the kindred knowledge society repositioning the relationship between state, market, and civil society. In this novel setting, education is indispensable in achieving the necessary human capital development and in adequately responding to the work and employment changing conditions. Education, thus, is expected to contribute to the development of business-oriented entrepreneurship, as it is eloquently argued in the discourse of many international initiatives, such as the World Economic Forum (2009) and international organizations, as is the European Union (EU):

> *Reinforcing entrepreneurial education in schools, vocational education institutions and universities will have a positive impact on the entrepreneurial dynamism of our economies.*
>
> *European Commission (2014: 1)*

A second set of factors refers to the increased awareness of economic inequalities, social exclusion, and environmental degradation. There is a widespread concern about the limits and imbalances of (neoliberal) globalization pinpointing the need for a more sustainable model of development, one that acknowledges and addresses the economic, social, and environmental challenges. In this context, understandings of entrepreneurship are mainly focused on creating novel solutions to global and local sustainability issues and social problems. Education, thus, becomes a powerful instrument for changing mindsets about what is sustainable development and how it can be achieved. "Entrepreneurship education," as an indispensable dimension of this endeavor, focuses on sustainability and social entrepreneurship, that is, contributes to Sustainable Development Goals (SDGs), the emblematic initiative of the United Nations Organization[1] (UN, 2015).

"Entrepreneurship education," therefore, is a common subject in two prominent discourses on educational reform: those that strongly support development of the "21st-century skills and competences" (P21, 2007) for economic development, and those that prioritize sustainable development, and consider entrepreneurship education to be an enabling factor of ethical, social, and human values, competences, and attitudes required for individual and collective actions. In both aspects, traditional educational approaches are criticized as inadequate in keeping up with change and in preparing students for the emerging economic and social necessities.

Another set of factors focuses on the form, content, and outcomes of entrepreneurship education initiatives. In this context, some scholars distinguish entrepreneurship from enterprise education: Jones and Iredale (2010: 11) note that

the primary focus of entrepreneurship education is on starting, growing, and managing a business, whereas that of enterprise education is on the acquisition and development of personal skills, abilities, and attributes that can be used in different contexts and throughout the life course.

In the same vein, Gibb (2005: 46) distinguishes between enterprising behavior and entrepreneurial behavior.

Indeed, there can be a "narrow" and a "broad" definition of entrepreneurship education. But, in our view, there is a significant common ground between the two approaches: First and foremost, they share the concern for developing competences required for students to improve their lives and create value for themselves and others. Second, many education programs combine features from both discourses. Entrepreneurship education can take the form of a learning method, a learning process, and a learning content (Pittaway and Cope, 2007); an active pedagogy, a flexible knowledge-generating program; a way of thinking, designing, and problem-solving (Amos and Onifade, 2013); a call for action; or an implemented practice. As such, it can be offered in many different contexts and forms and serve more than one purpose.

Therefore, entrepreneurship education (either business oriented or social and sustainable development oriented, "broad" or "narrow") is justifiably included in the programs of study and the curricula of literally every level and type of education, from early childhood education and care throughout school and university education, vocational training, and adult learning. In this chapter, we discuss a course within the initial early childhood teacher education (TE), containing elements of both entrepreneurship and enterprise education and focusing on social entrepreneurship. The course's focus, form, method, content, and expected outcomes echo our firm belief that entrepreneurship expands beyond business creation, maintenance, and development, towards nurturing an entrepreneurial mindset, enabling future teachers to identify creative ideas and turn them into actions.

---

[1] Entrepreneurship education is within the scope of initiatives for achieving SDG 3 ("Ensure healthy lives and promote well-being for all at all ages"), SDG 4 ("Ensure inclusive and quality education for all and promote lifelong learning"), and SDG 8 ("Promote inclusive and sustainable economic growth, employment, and decent work for all").

In the sections that follow, first, we briefly review social entrepreneurship within the context of sustainable development, and then we discuss the case study of teaching social entrepreneurship to students at the National and Kapodistrian University of Athens.

7.2   CONCEPTUAL FRAMEWORK

7.2.1   Social Entrepreneurship for Sustainable Development

Sustainable development is undoubtedly one of the most challenging and comprehensive discourses of our time, demanding action from international organizations and corporations, governments, markets, public and private enterprises, the civil society, and individual citizens. Sustainable development and entrepreneurship converge in the newly established field of "sustainable entrepreneurship" (Shepherd and Patzelt, 2011). In their widely used book on entrepreneurship, Hisrich, Peters, and Shepherd (2017: 22) argue for the close links between the two concepts: "Without entrepreneurial knowledge, opportunities for sustainable development are unlikely to become a reality."

The significance of entrepreneurship in bringing about change towards sustainability has been debated since the 1990s. The metaphor of the "triple bottom line"[2] (TBL, Elkington, 1998) has inspired not only measurement indicators, monitoring, and management practices, but also influenced changes in entrepreneurship education, in the framework of Corporate Social Responsibility programs (Schmidpeter, 2014). Savitz (2014: 53) notes that the TBL and "sustainability movement" urged businesses to address as a whole several areas that were previously dealt with "in isolation from one another," that is, "the environment, community relations, labor practices, responsible investment, and others," including the development of educational initiatives.

Several UN initiatives have reiterated "the centrality of entrepreneurship" for achieving the SDGs (especially those of SDG 4 and SDG 8 mentioned in the previous section) and have argued for "the contribution of entrepreneurship to poverty eradication, employment, and economic empowerment," and its potential "for the integration of vulnerable groups" (UNCTAD, 2017: 1–2).

On a rather different key, many scholars question the potential of entrepreneurship for societal transformation and the creation of sustainable economies (Hall, Daneke and Lenox, 2010). Some scholars find that the links between entrepreneurship and sustainability are unclear, while recognizing that in developing countries, entrepreneurship is compelled to unsustainable behavior to maximize profit (Dhahri and Omri, 2019).

Recent scholarly publications thoroughly examine every aspect of this important relationship (Apostolopoulos et al., 2018; Markman et al, 2016; Shepherd and Patzelt, 2017; Muñoz and Cohen, 2018). It is clear from this body of literature that even the

---

[2] The concept of "TBL" refers to the performance levels and future targets for economic prosperity, environmental quality, and social justice, against which businesses and economies should be held accountable in view of sustainable development.

skeptics assume that various forms of entrepreneurship can have an impact on many economic, social, and environmental challenges at the global and local levels. It is also evident that an increasing number of entrepreneurship scholars, researchers, and practitioners understand the potential and impact of entrepreneurial action beyond economic growth. Thompson, Kiefer, and York differentiate between three groups of people who share these concerns: "social entrepreneurs," who "focus mainly on problems that affect *people today*"; "sustainable entrepreneurs," who "focus on a 'TBL' of *people, planet, and profit*"; and "environmental entrepreneurs," who "are focused on creating simultaneous *economic and ecological* benefit" (Thompson, Kiefer and York, 2011: 204, emphasis in the original).

This chapter focuses mostly on social entrepreneurship, which for some scholars also refers to environmental challenges (i.e., Chahine, 2016: 1–2). There is a wide spectrum of definitions of social entrepreneurship. According to Dees, one of the pioneers in the field, the concept of social entrepreneurship refers to "entrepreneurial approaches to social problems," which combine "the passion of a social mission with an image of business-like discipline, innovation, and determination" (Dees, 2001: 1). He describes social entrepreneurs as "change agents in the social sector," who operate by

Adopting a mission to create and sustain social value (not just private value).
Recognizing and relentlessly pursuing new opportunities to serve that mission.
Engaging in a process of continuous innovation, adaptation, and learning.
Acting boldly without being limited by resources currently in hand.
Exhibiting heightened accountability to the constituencies served and for the outcomes created

(Dees, 2001: 4).

For Nicholls (2006: 22), social entrepreneurship refers to "innovative and effective activities that focus strategically on resolving social market failures and creating new opportunities to add social value systemically by using a range of resources and organizational formats to maximize social impact and bring about change." Austin, Stevenson, and Wei-Skillern (2006: 371) note that this "social value creating activity... can occur within or across the nonprofit, business, or government sectors."

Social entrepreneurship is currently exercised within a vast array of economic, educational, research, welfare, social, and spiritual activities, the majority of which are in the domain of nongovernment, not-for-profit organizations (Bornstein and Davies, 2010). It offers innovative, problem-solving approaches in sectors that the state has proven "ill-equipped to deal with many of the modern social problems condemning people in a state of dependency and poverty" (Leadbeater, 1997: 1–2). It makes effective use of limited resources and acts productively even within institutional constraints (Desa, 2012). Moreover, it is particularly useful in conditions of economic, social, and even humanitarian crises, setting in motion "a virtuous circle of social capital accumulation," which gives communities "a better chance of standing on their own two feet" (Leadbeater, 1997: 3).

Nobelist Muhammad Yunus contends that "social entrepreneurship is an integral part of human history" and calls social entrepreneurs "do-gooders" (Yunus, 2006: 40).

Focusing on those who manage to operate their social enterprises in "more than full cost recovery," he names these persons or groups "social business entrepreneurs." He asserts that they can become "very powerful players in national and international economy" and indeed "broaden the concept of the market" and the "interpretation of capitalism." "Once this is done," he predicts "social business entrepreneurs can flood the market and make it work for social goals as efficiently as it does for personal goals" (2006: 41–42).

In our view, a course for social entrepreneurship to students of higher education, within the context of a crisis-hit European economy, does exactly that: It broadens the interpretation of capitalism and offers these young people more choices for achieving their objectives.

7.2.2 ENTREPRENEURSHIP IN HIGHER EDUCATION: BUSINESS, SOCIAL, AND SUSTAINABLE ENTREPRENEURSHIP

The literature on entrepreneurship education in HEIs, and research on the content, modes of delivery, and impact, has grown significantly during the past 15 years (EC, 2008; Weber, 2012; Fayolle and Gailly, 2015; Volkmann and Audretsch, 2017). Referring to the content and form of delivery, Pittaway and Edwards (2012) argue that entrepreneurship education courses and programs mainly take one of the following forms: learning "about" entrepreneurship, with emphasis on raising awareness, offering information, and developing students' relevant knowledge; learning "for" entrepreneurship aiming at counseling and enabling students to acquire relevant competences and key skills; and learning "through" entrepreneurship, where students are offered the opportunity for "hands-on" experiences, by developing or, even further, materializing businesses. However, the evidence is unclear as to which approach is working better or which model shows more potential in impacting students.

As it pertains to social entrepreneurship, Thiru (2011), having studied several university curricula, argues that they may be classified into three approaches: "accommodation" when "social enterprise is included in the curriculum in the form of one or more courses within the existing degree programs as electives only"; "integration" when institutions "use specific social enterprise-related programs in their curriculum, most of them with strong cocurricular offerings"; and "immersion" when they "integrate social enterprise curriculum in existing programs, focusing strongly on field study and collaboration with social enterprise practitioners" (Thiru, 2011: 186–187).

The UN Higher Education Sustainability Initiative (HESI, 2017), the European Commission-funded project "University Educators for Sustainable Development" (UE4SD, 2014), and the EU and OECD joint platform "HEInnovate" (EC, 2015), among other initiatives, have shown that higher education institutions can play a crucial role in supporting sustainable development programs, and in many cases, they do so by providing education and activities in social entrepreneurship. European universities, in particular, show a rich record of entrepreneurship education initiatives, implying that entrepreneurship "is also about creativity, innovation, and growth, a way of thinking and acting relevant to all parts of the economy and society as well as the whole surrounding ecosystem" (Volkmann and Audretsch, 2017: 3).

Entrepreneurship education courses can also be found in many TE programs, especially in the EU, where official recommendations, initiatives, and scholarly research promote and support its development (GHK, 2011).

Courses on entrepreneurship within the Greek universities have existed since the mid-2000s. Many departments have been very ambivalent towards entrepreneurship for a long time, and the introduction of such courses in noneconomics departments has been a source of controversy, as many students and staff identified them with an alleged " backdoor" towards the "entrepreneurial university" and a "surrender" of Greek universities to the "commanding calls" of neoliberalism (i.e., Gounari, 2012; Sotiris, 2012). These courses have been subsidized, in the main, by the EU,[3] thus contributing to concerns by several parties that a neoliberal agenda is introduced in higher education (Gouvias, 2011).

The development of entrepreneurship courses is supported by the Units for Innovation and Entrepreneurship of the Greek universities (UIE/Greek acronym: MOKE). These units, within every Greek HEI, have used subsidies from the EU Operational Programs for Education (OECD, 2017). Their mandate includes providing support for the development of undergraduate and postgraduate courses, student competitions, conferences, and other activities, aim at the development of entrepreneurship knowledge, skills, and competences, as well as the cultivation of students' entrepreneurial mindset. They also act as intermediaries, seeking potential investors for students' business proposals. UIEs work closely with the University Structures for Employment and Career development, the Liaison Offices, and the Offices for Field Practice of Students to develop an interface between Greek higher education and the labor market. The cultivation of entrepreneurial spirit among students encourages them to entrepreneurial attempts, contrary to widespread views against entrepreneurship. They usually adopt learning models "about" and "for" entrepreneurship that include lectures and seminars on innovation and entrepreneurship, study visits to businesses, preparation of business plans based on students' ideas, and university-wide student competitions on entrepreneurship. The UIE supports departments that include entrepreneurship courses in their curricula by assigning tutors and business plan consultants and providing funds for guest lectures and field visits. Until the mid-2010s, most entrepreneurship courses and programs in Greek universities focused on business entrepreneurship, as the sector of social entrepreneurship was still small.[4] The crisis played a role in raising awareness for these activities and pointing to the need for relevant courses, as there have been many voices praising the role of social entrepreneurship in leaving the crisis behind (i.e., Liargovas and Apostolopoulos, 2017).

---

[3] This is, by European Structural Funds through the second Operational Program for Education and Initial Vocational Training (OPEIVT II, Greek acronym: EPEAEK II) and subsequently by the Operational Program for Education and Lifelong Learning (OPELLL9, Greek acronym: EPEDVM). For a detailed description of the proposed policy measures and operational procedures, see EPEAEK II (2008).

[4] Social entrepreneurship experienced significant growth after the 2008–2009 recession. In Greece, it involved 1.9% of the active population at that time, around the average of GEM countries (GEM, 2009), despite lacking a legal framework (the first law for social entrepreneurship was enacted in 2011, Law 4019/2011). Social entrepreneurship still remains limited in Greece, in comparison with other European countries, but the visibility of enterprises in this sector has increased considerably (GEM, 2015).

Business-oriented entrepreneurship remains at the core of the majority of courses offered in Greek universities. There is research evidence showing that a large percentage of students still do not conceive entrepreneurship as a collective endeavor benefitting society, something that could have an impact on choosing "social entrepreneurship in the field of entrepreneurial action" (Papagiannis, 2018: 16). Meanwhile, the growing number of initiatives for the development of social entrepreneurship in Greece, such as Impact Hub Athens, HIGGS incubator, "Think Social Act Business" training program of British Council, and "Social Develop Athens" program of the Municipality of Athens, raise the attention of social alerted students to learn how to implement their ideas for solving crucial social problems, which have been enlarged during the period of crisis in Greece. Therefore, despite the prejudice that education and entrepreneurship should not overlap, social entrepreneurship education seems to be the way for a social and sustainable development (Jensen, 2014).

7.3 THE CASE STUDY OF TEACHING SOCIAL ENTREPRENEURSHIP FOR SUSTAINABLE DEVELOPMENT TO EARLY CHILDHOOD EDUCATION STUDENTS AT THE UNIVERSITY OF ATHENS

7.3.1 THE COURSE ON "EDUCATION FOR INNOVATION AND DEVELOPMENT IN MODERN GREEK SOCIETY": FROM BUSINESS TO SOCIAL ENTREPRENEURSHIP

The course on "Education for Innovation and Development in Modern Greek Society" was introduced at the Department of Early Childhood Education (DECE) at the National and Kapodistrian University of Athens in 2011. DECE was no exception as concerns the reservations towards entrepreneurship education. Actually, it required a certain amount of dialoguing in order to convince those reacting, that such an endeavor should not be identified with neoliberal ideas but rather viewed as an attempt to develop a course aiming at helping students express and develop their ideas and consequently turn them into action, and at adding to their studies economic value and personal fulfillment. The title of the course, "Education for Creativity and Development in Modern Greek Society," reflects the way the course tutors (and authors of this chapter) view "entrepreneurship," that is, an essential skill for future teachers, along the lines of the definitions given in the previous sections of this chapter. As stated in the Study Guide, the course aims at

> *developing a critical understanding of the ways the globalized economy of knowledge has influenced societies of late modernity... at cultivating skills that are cardinal in our contemporary world such as skills in decision making and taking initiatives, in entrepreneurship and transforming one's ideas into action. These skills are intricately connected with creativity, innovative thinking, and risk-taking, as well as the ability to plan forward for achieving particular goals in one's private, social, and work life.*[5]

The course is offered to students in their final fourth year of study. It is an elective, research-oriented course belonging to the "Research Practice" cycle, offering up to

---

[5] www.ecd.uoa.gr/wp-content/uploads/2013/06/StudentGuide2018-19.pdf, p. 134.

25 students places in its roaster.[6] Students have to take one out of seven such courses and develop their own research project. Successful completion of their research project is mandatory in obtaining their undergraduate degree. If we were to use Thiru's (2011) classification, the course falls into what he calls the "accommodation" approach since it is included in the curriculum within the existing degree program as an elective.

Course development was informed by the methodological principles of action research; that is, we were open to improving our practice based on the cyclical process of reflecting on practice, taking action, reflecting again, and taking further action (Reason and Bradbury, 2007). The team of instructors systematically improved their design on the basis of what is relevant to students and makes a real difference in their lives.

A big challenge in embarking upon the course in 2011 was to reverse the reservations held by the departments' instructors and the student body. It was no surprise to us to see that doubts gradually evaporated even more so when the economic crisis culminated and "entrepreneurial spirit" and skills gained attention as potential solutions for overcoming problems of graduates' unemployment or as pathways towards more inspiring and interesting careers, outside standard teachers' employment options. The course retained its positive momentum as the difficult years were left behind and Greece returned to economic normality.

In the first 5 years, the course focused on business entrepreneurship, following, to a great extent, the curriculum content proposed by the University of Athens UIE. The business-planning exercise had been the core activity of our course, as students' business plans were the required deliverables for their assessment. The usefulness of business planning, along with criticisms (i.e., because of their restrictive structure), is well documented in the literature of entrepreneurship education (i.e., Fayolle, 2007; van der Sijde et al., 2008). The business plan model we adopted followed the format proposed by the UIE (UOA-UIE, 2013: 26–49) and required increased knowledge of economics, something that raised the degree of difficulty of the course for our (noneconomics) students. It required activities that future teachers are not familiar with, including market sector research; risk and future growth assessment; competitor analysis; planning for production, marketing, and human resources; and financial planning and budgeting.

The students developed their ideas into business plans with the help of the tutors and investing much more time and effort than was usually required for other courses in their program of studies. Their enterprise ideas involved mostly small-scale businesses that were relevant to their training as future educators of young children (play spaces; children's books publishing; early childhood care and education centers; in-service training programs; and so forth). Although the course was oriented towards business entrepreneurship, we included a fair share of knowledge and discussion about pressing social issues. Thus, we focused on creativity and innovation informed by new theoretical approaches and practices of entrepreneurship within the realm of

---

[6] One distinctive feature of the student population of the DECE is that it is predominantly female. In the past two academic years (2017–2018 and 2018–2019), the course was taken by female students exclusively.

social and sustainable development sectors. Many of the students—sensitive about social issues, especially those involving children—developed business plans that showed increased social awareness and preference for social enterprises. Every year at least one of these business plans gained remarkable attention at the university-wide competitions, and achieved awards outbidding business plans from teams coming from departments offering economic-oriented studies.

The shift towards social entrepreneurship took place in 2016 at the peak of the Greek financial crisis. Under circumstances of scarce bank funding, high taxation, huge unemployment reaching at the time 48.3% for ages 15–24 (see ECB, 2017), and deep social crisis, but at the same time increased community and solidarity initiatives, we felt that a course on social entrepreneurship and sustainable development would help our students to develop both their entrepreneurial and enterprise competences and broaden their horizons as to what an early childhood educator can do besides standard teachers' employment options.

Reflecting on our practices and capitalizing on students' feedback, we gradually limited theory and lectures, giving more room for tutorials and field visits. Social entrepreneurship provided us with a coherent scaffolding of knowledge, practices, and experiences, within which our students could see themselves and develop their ideas.

We also changed the business-planning exercise introducing the much more-friendly "Business Model Canvas (BMC)" (Osterwalder and Pigneur, 2010). The BMC consists of nine sections and can be looked at as a conceptual map. The inventors of the model claim that it is created to help develop strategies and plans, and make decisions around the different elements of a business. We found this model to be more suitable for students of limited knowledge in economics. It provides a basic structure for brainstorming about new ideas and approaches; it does not require knowledge of the business jargon and enhances collaborative work between the students. We also agree with Stenn (2017: 60), who sees in BMC "a design-thinking tool developed to create more access to business thinking, planning, strategizing, and problem-solving."

The business plans of our students included, among others, the development of "Social Cooperative Enterprises" aiming at helping, educating, and providing inclusive environments for refugee children and children with special education needs and disabilities, a cooperative art-centered playground, a counseling service for parents and children, a healthy food catering for school canteens that would also cater for homeless people, and an ecological awareness product and services center.

7.3.2 THE PEDAGOGY ADOPTED

Right from the beginning, the course was designed in a multimodal fashion. It is taught jointly by three instructors and not by one, as is usually the case, a social psychologist, a pedagogue focused on educational planning and pedagogical practices, and an economist focused on economics education. It is a collaborative effort, where every member of the teaching team brings in her/his own perspective and expertise within a commonly agreed framework. Moreover, the course that lasts 13 weeks includes lectures, individual and group tutorials assisting students in developing their business plan, visits to local enterprises, and invited lectures, seminars, and workshops by experts and entrepreneurs. The students work in small groups of three or four, and a business plan, based on an original entrepreneurial idea, is the final deliverable for each group.

Our approach challenges the traditional model whereby the teacher's task is to impart knowledge or skills to students and moves away from teacher initiating and controlling interaction. We assume the role of facilitators supporting the process of learning. We share our knowledge enabling students to reach goals in which they are personally invested. As facilitators, we involve students in establishing topics for exploration, locating relevant resources, and setting up conditions of dialogue so as learners can design their own business plan. Our teaching promotes experiential learning and the articulation between theory and praxis; it cultivates agency and collaborative work. It is founded on the principle that learning is relational, takes place in social environments where collaborative activities provide learners with opportunities to communicate, interact, and consequently construct their own world of knowledge. In other words, we move from the monologic to the dialogic classroom (Gergen, 2015).

Moreover, the course moves beyond the walls of the university in an attempt to reduce the barrier between the inside and the outside. We try to bring together students, instructors, and entrepreneurs in the belief that learning occurs most effectively through the interested participation of all parties involved in social entrepreneurship. As Gergen (ibid: 157) would claim, in a social constructionist vein, we build enthusiasm together. To give an example of the 2018–2019 program, students met inside or outside the classroom in specific social enterprise venues, with various social entrepreneurs active in organizing nonformal education projects, cultural and nature tours, creative children's activities, anti-bullying programs; in training guide dogs for the blind; in providing catering services; and in setting up a multipurpose café employing persons with disabilities and other similar projects. All such initiatives aimed at promoting education, culture, diversity, and inclusion and act or support social change. Students also had the opportunity to meet with people active in social entrepreneurship from a different angle and different academic fields. Thus, we were visited by experts from the social entrepreneurship unit of Athens University of Economics and Business, social entrepreneurship businessman, and social entrepreneurship legal consultants. This multimodal approach steps away from the traditional classroom towards a vibrant learning context, thus contributing to building a learning community.

Coaching in tutorials attends to students' aspirations, discusses their progress as their ideas develop, offers affirmation, and provides a sense of confidence in their skills. We have used business planning as a pedagogical tool, valuing the opportunities this exercise gives our students to think, make decisions, and act as entrepreneurs; to understand the business environment; and to develop their programming, analytic, synthetic, communicative, entrepreneurial, and enterprising skills and capabilities.

7.4 RESULTS AND DISCUSSION: EVALUATION AND ASSESSMENT OF THE COURSE'S IMPACT

As regards the evaluation of student performance, we try to move away from the traditional assessment method of testing. In place of examinations, we evaluate our students' business plans that are a collaborative output built on regular feedback provided in the tutorials. As a result, evaluation is formative rather than summative.

At the end of the term, students present their business plans outside the classroom in the friendly environment of an open to cultural events bookstore near the university. They discuss their projects in front of their fellow students and a colleague from the Athens University of Economics and Business who provides critical reflections. Thus, assessment is relationally based and becomes a "dialogic evaluation" (Green, 2001; McNamee, 2015) in a participatory, engaged, inclusive, and respectful interaction that is more of a celebration marking the end of a process rather than a success or failure terrifying event. Students perceive the knowledge they have acquired as a relational achievement and not as a private cognitive process.

On the same occasion, they are engaged reflexively in what they have achieved. We quote some of the reflections voiced by students in the closing event of the academic year 2018–2019:

> *This course made you feel confident in yourself...that you can actually make it...that there is a solution to everything.*
>
> *It is the first course that we found ourselves in a situation where we could defend our own idea.*
>
> *I felt psychologically exalted to the here and now... to what we experience as human beings... we can achieve anything collaboratively and collectively.*
>
> *My idea was enriched... I rest assured... as long as keep your eyes open and your goal is the human being.*

It is clear from the above comments that students blossom when what they learn can be connected productively to their own lives.

Besides this evaluative process, we ask students to respond to an open questionnaire, as part of the course evaluation required by the university. In their answers, students use a less spontaneous discourse and followed the tradition of "objective" evaluation residing in the mind of the individual student. Clearly, this is not a relational exercise. Analysis of the students' answers in the 2018–2019 academic year reveals overall similarities in their responses: Almost all underlined the value of working collaboratively in small groups; they acknowledged that this is not always an easy task, but it is very rewarding when achieved. The importance of decision-making, risk taking, creativity, and innovation was highlighted. Many found it difficult to grasp unfamiliar economic concepts but they enjoyed putting theory and action together. The overall spirit was that of a sense of accomplishment that made students proud of themselves.

We quote some answers given:

> *We went beyond the traditional bounds... The entire process was an effort on my part to "say" what I want.*
>
> *This course is like no other.*
>
> *I thought I hated business and finance but now I want to practice this knowledge in connection with what I study.*
>
> *With this course I saw how I can open doors. A lot of work and perseverance is required in order for me to see which is the door I want to open. But without risk taking I will never find out.*
>
> *A business plan requires kindness and boldness; respect towards our fellow men and our colleagues.*

The business plan presentations and the students' evaluation of the course echo Volkmann's and Audretsch's (2017: 3) claim that social entrepreneurship education "is not only about creating business plans but also about a new way of thinking." Students underlined the importance of creativity and innovation, while they seem to have learnt how to mobilize collaborative processes in the service of effective action. In designing their business plans, they made decisions that carried the voices of all the members of their group. We believe that the element of social mission, inherent in social entrepreneurship as noted by Dees (2001), is responsible for the enthusiasm prevalent in students' evaluation of the course. In all our discussions, these young women were very impressed with the social entrepreneurs they came in touch with; they were inspired by the social impact of their social businesses and the potential for social change. Bornstein's and Davies' (2010) assertion that social entrepreneurs embolden others to pursue similar ideas and solutions has been at work in our case.

7.5 CONCLUDING REMARKS

Entrepreneurship education is considered one of the basic elements for achieving the relevance of education systems to 21st-century needs and challenges, including every aspect of sustainable development. Independently of their focus, business, social and environmental entrepreneurship programs, and courses create value and foster closer relationships between students and teachers of all levels of education with experts, practitioners of the market, and active members of civil society. In this chapter, we discussed the development, implementation, and assessment of entrepreneurship education at the university level, in the DECE at the University of Athens. We combined forms of delivery, content, pedagogy, and expected learning outcomes, to focus our course on social entrepreneurship, aiming at nurturing the entrepreneurial mindset, the competences and skills, that would enable our students to turn their creative ideas into actions with social value. In the 9 years, we have been teaching this course we have experienced a transformative process with students growing and ourselves, questioning our assumptions and equally changing in a meaning-making, relational context. We have witnessed the generative possibility of entrepreneurship education.

REFERENCES

Amos, A. and Onifade, C. A. (2013). The perception of students on the need for entrepreneurship education in teacher education programme. *Global Journal of Human-Social Science Research*, 13(3), 75–80.

Apostolopoulos, N., Al-Dajani, H., Holt, D., Jones, P. and Newbery, R. (2018). Entrepreneurship and the sustainable development goals. *Contemporary Issues in Entrepreneurship Research*, 8, 1–7.

Auerswald, P. (2012). *The Coming Prosperity: How Entrepreneurs are Transforming the Global Economy*. New York: Oxford University Press.

Austin, J. E., Stevenson, H. and Wei-Skillern, J. (2006). Social and commercial entrepreneurship: Same, different, or both? *Entrepreneurship Theory and Practice*, 30, 1–22.

Bauman, Z. (2005). *Work, Consumerism and the New Poor*, 2nd edition. London: Open University Press.

Beck, U. (1992). *Risk Society*. London: SAGE Publications.

Besley, T. and Peters, M. A. (2007). *Subjectivity and Truth: Foucault, Education, and the Culture of Self*. New York: Peter Lang.

Bornstein, D. and Davies, S. (2010). *Social Entrepreneurship: What Everyone Needs to Know*. New York: Oxford University Press.

Chahine, T. (2016). *Introduction to Social Entrepreneurship*. Boca Raton, FL: CRC Press.

Dees J. G. (2001). The meaning of "social entrepreneurship". Duke University, Durham. Available online: https://centers.fuqua.duke.edu/case/wp-content/uploads/sites/7/2015/03/Article_Dees_MeaningofSocialEntrepreneurship_2001.pdf.

Desa, G. (2012). Resource mobilization in international social entrepreneurship: Bricolage as a mechanism of institutional transformation. *Entrepreneurship Theory and Practice*, 36(4), 727–751.

Dhahri, S. and Omri, A. (2019). Entrepreneurship contribution to the three pillars of sustainable development: What does the evidence really say? *World Development*, 106(C), 64–77.

Drori, G. S., Jang, Y. S. and Meyer, J. W. (2006). Sources of rationalized governance: Crossnational longitudinal analyses, 1985–2002. *Administrative Science Quarterly*, 51, 205–229.

Elkington, J. (1998). *Cannibals with Forks: The Triple Bottom Line of 21st Century Business*. Oxford: New Society Publishers.

EPEAEK II. (2008). *Supporting Entrepreneurship Ideas in Higher Education*. Athens: Acronym. Available online: http://repository.edulll.gr/150.

European Central Bank. (2017). *Recent Developments in Youth Unemployment*. Frankfurt: ECB. Available online: www.ecb.europa.eu/pub/pdf/other/ebbox201703_02.en.pdf.

European Commission. (2007). *Key Competences for Lifelong Learning*. Brussels: European Commission, p. 104.

European Commission. (2008). *Final Report of the Expert Group: Entrepreneurship in Higher Education, Especially within Non-Business Studies*. Brussels: European Commission Directorate - General for Enterprise and Industry.

European Commission - Entrepreneurship 2020 Unit. (2014). *Entrepreneurship Education: A Guide for Educators*. Brussels: European Commission Directorate - General for Enterprise and Industry.

European Commission. (2015). *Entrepreneurship Education: A Road to Success: A Compilation of Evidence on the Impact of Entrepreneurship Education Strategies and Measures*. Brussels: European Commission.

Fayolle, A. (ed.) (2007). *Handbook of Research in Entrepreneurship Education: A General Perspective* (Vol. 1). Cheltenham: Edward Elgar Publishing.

Fayolle, A. and Gailly, B. (2015). The impact of entrepreneurship education on entrepreneurial attitudes and intention: Hysteresis and persistence. *Journal of Small Business Management*, 53(1), 75–93.

GHK. (2011). *Mapping of Teachers' Preparation for Entrepreneurship Education*. Brussels: DG Education and Culture.

Gibb, A. (2005). The future of entrepreneurship education: Determining the basis for coherent policy and practice? In: Kyrö, P. and Carrier, C. (eds.) *The Dynamics of Learning Entrepreneurship in a Cross-Cultural University Context*, Entrepreneurship Education Series 2/2005 (pp. 44–67). Hämeenlinna: University of Tampere, Research Centre for Vocational and Professional Education.

Global Entrepreneurship Monitor – GEM. (2009). 2009 global report. Global Entrepreneurship Research Association. Available online: www.gemconsortium.org/report/gem-2009-global-report.

Global Entrepreneurship Monitor. (2015). Special topic report: Social entrepreneurship. Global Entrepreneurship Research Association. Available online: www.gemconsortium.org/report/49542.

Gounari, P. (2012). Neoliberalizing higher education in Greece: New laws, old free-market tricks. *Power and Education*, 4(3), 277–288.

Gouvias, D. (2011). EU funding and issues of "marketisation" of higher education in Greece. *European Educational Research Journal*, 10(3), 393–406.

Green, J. C. (2001). Dialogue in evaluation: A relational perspective. *Evaluation*, 7, 181–187.

Gergen, K. (2015). *An Invitation to Social Construction*, 3rd edition. London: Sage.

Hall, J. K., Daneke, G. A. and Lenox, M. J. (2010). Sustainable development and entrepreneurship: Past contributions and future directions. *Journal of Business Venturing*, 25(5), 439–448.

Higher Education Sustainability Initiative - HESI. (2017). United Nations higher education sustainability initiative. Available online: https://sustainabledevelopment.un.org/sdinaction/hesi.

Hisrich, R. D., Peters, M. P., Shepherd and D. A. (2017). *Entrepreneurship*, 10th edition. New York: McGraw-Hill Education.

Jensen, T. L. (2014). A holistic person perspective in measuring entrepreneurship education impact: Social entrepreneurship education at the Humanities. *The International Journal of Management Education*, 12(3), 349–364.

Jones, B. and Iredale, N. (2010). Enterprise education as pedagogy. *Education + Training*, 52(1), 7–19.

Leadbeater, Ch. (1997). *The Rise of the Social Entrepreneur (Demos Papers)*. London: Demos.

Liargovas, P. and Apostolopoulos, N. (2017). Opinion: Social entrepreneurship, the antidote for crisis. *Kathimerini*, 24/04/2017. [in Greek]. Available online: www.kathimerini.gr/906276/article/oikonomia/ellhnikh-oikonomia/apoyh-koinwnikh-epixeirhmatikothta-to-antidoto-ths-krishs.

Markman, G. D., Russo, M., Lumpkin, G. T., Jennings, P. D. and Mair, J. (2016). Entrepreneurship as a platform for pursuing multiple goals: A special issue on sustainability, ethics, and entrepreneurship. *Journal of Management Studies*, 53(5), 673–694.

McNamee, S. (2015). Evaluation in a relational key. In: Dragonas, T., Gergen, K., McNamee, S. and Tseliou, E. (eds.) *Education as Social Construction: Contributions to Theory, Research and Practice* (pp. 336–349). Chagrin Falls, OH: Taos Institute Publications/ WorldShare Books.

Muñoz, P. and Cohen, B. (2018). Sustainable entrepreneurship research: Taking stock and looking ahead. *Business Strategy and The Environment*, 27, 300–322.

Nicholls, A. (ed.) (2006). *Social Entrepreneurship: New Models of Sustainable Social Change*. New York: Oxford University Press.

OECD. (2017). *Inclusive Entrepreneurship Policies, Country Assessment Notes: Greece, 2017*. Paris: OECD.

Osterwalder, A. and Pigneur, Y. (2010). *Business Model Generation: A Handbook for Visionaries, Game Changers, and Challengers*. Hoboken, NJ: John Wiley & Sons.

P21 – Partnership for 21st Century Learning. (2007). Framework for 21st century learning. Available online: http://static.battelleforkids.org/documents/p21/P21_framework_0816_2pgs.pdf.

Papagiannis, G. (2018). Entrepreneurship education programs: The contribution of courses, seminars and competitions to entrepreneurial activity decision and to entrepreneurial spirit and mindset of young people in Greece. *Journal of Entrepreneurship Education*, 21(1), 1–21.

Pittaway, L. and Edwards, C. (2012). Assessment examining practice in entrepreneurship education. *Education and Training*, 54, 778–800.

Pittaway, L. and Cope, J. (2007). Entrepreneurship education: A systematic review of the evidence. *International Small Business Journal*, 25(5), 479–510.

Reason, P. and Bradbury, H. (2007). *Handbook of Action Research*, 2nd edition. London: Sage.

Rose, N. and Osborne, Th. (2000). Governing cities, governing citizens. In: Isin, E (ed.) *Democracy, Citizenship and the City: Rights to the Global City* (pp. 95–109). London: Routledge.

Savitz, A. and Weber, K. (2014). *The Triple Bottom Line*. San Francisco, CA: Jossey-Bass.

Shepherd, D. A. and Patzelt, H. (2011). The new field of sustainable entrepreneurship: Studying entrepreneurial action linking "what is to be sustained" with "what is to be developed". *Entrepreneurship Theory and Practice*, 35(1), 137–163.

Shepherd D. A. and Patzelt, H. (2017). Researching entrepreneurships' role in sustainable development. In: Shepherd, D. A. and Patzelt, H. (eds.) *Trailblazing in Entrepreneurship* (pp. 149–179). Cham: Palgrave Macmillan.

Schmidpeter, R. (2014). The evolution of CSR from compliance to sustainable entrepreneurship. In: Weidinger, Ch., Fischler, F. and Schmidpter, R. (eds.) *Sustainable Entrepreneurship* (pp. 127–134). Heidelberg: Springer.

Sotiris, P. (2012). Theorizing the Entrepreneurial University: Open questions and possible answers. *Journal of Critical Education Policy Studies*, 10(1), 112–126.

Stenn, T. L. (2017). *Social Entrepreneurship as Sustainable Development Introducing the Sustainability Lens*. Cham: Palgrave Macmillan.

Thiru, Y. (2011). Social enterprise education: New economics or a platypus? In: Lumpkin, G. Th. and Katz, J. A. (eds.) *Social and Sustainable Entrepreneurship* (pp. 175–200). Bingle: Emerald Group Publishing Limited.

Thompson, N. A., Kiefer, K. and York, J. (2011). Distinctions not dichotomies: Exploring social, sustainable, and environmental entrepreneurship. In: Katz, J. and Corbett, A. C. (eds.) *Social and Sustainable Entrepreneurship* (pp. 201–229). Bingley: Emerald Books.

UE4SD. (2014). University educators for sustainable development. European Union. Available online: www.ue4sd.eu/.

United Nations. (2015). Transforming our world: The 2030 agenda for sustainable development. New York: United Nations, Department of Economic and Social Affairs.

UNCTAD. (2017). Promoting entrepreneurship for sustainable development: A selection of business cases from the empretec network (UNCTAD/DIAE/ED/2017/6). Available online: https://unctad.org/en/PublicationsLibrary/diaeed2017d6_en.pdf.

UOA-UIE (University of Athens – Unit for Innovation and Entrepreneurship) (2013). Training of trainers notes. Athens: University of Athens. Available online: http://repository.edulll.gr/edulll/handle/10795/2579 [in Greek].

van der Sijde, P., Ridder, A., Blaauw, G. and Diensberg, C. (2008). *Teaching Entrepreneurship: Cases for Education and Training*. Heidelberg: Physica-Verlag.

Vavrus, F. (2004). The referential web: Externalization beyond education in Tanzania. In: Steiner-Khamsi, G. (ed.) *The Global Politics of Educational Borrowing and Lending* (pp. 141–153). New York: Teachers College Press.

Volkmann, C. K. and Audretsch, D. B. (2017). *Entrepreneurship Education at Universities: Learning from Twenty European Cases*. Berlin: Springer.

Wagner, T. and Compton, R. A. (2012). *Creating Innovators: The Making of Young People Who Will Change the World*. New York: Scribner.

Weber, R. (2012). *Evaluating Entrepreneurship Education*. Wisenbaden: Springer Gabler.

World Economic Forum. (2009). *Educating the Next Wave of Entrepreneurs*. Geneva: World Economic Forum.

Yunus, M. (2006). Social business entrepreneurs are the solution. In: Nicholls, A. (ed.) *Social Entrepreneurship: New Models of Sustainable Social Change* (pp. 39–44). New York: Oxford University Press.

8 Developing an Innovative Pedagogy for Sustainability in Higher Education

Athanassios Androutsos and Vasiliki Brinia
Athens University of Economics and Business

CONTENTS

> Give a man a fish and you feed him for a day, teach a man to fish and you feed him for a lifetime.
>
> *Chinese Proverb*

8.1 INTRODUCTION

The current educational system has still not developed a standard pedagogical approach to provide the 21st-century skills and competencies for addressing the emerging future issues in the area of sustainability. A new pedagogical approach must be considered in order to bridge the gap between the current educational system and real-life needs.

A new culture of teaching and learning should take place in the area of sustainable development that will cultivate the future competencies, practices, and culture that are necessary for future employability opportunities in the wide sector of sustainability. Skills, competencies, and practices related to sustainable development include innovative skills for tackling with the complicated nature of the real-world problems, collaborative skills for dealing with global and interconnected issues, digital competencies for tackling with technology and innovation, art-based practices for cultivating and promoting creativity, and a co-creative mindset for designing human-centered solutions.

In this chapter, a new pedagogy will be proposed and evaluated by using approaches based on a designedly way of thinking, digital economy, art-based practices, and entrepreneurial spirit. Design thinking in higher education will lead students – through creative and artistic methods – to be much more innovative, and also will create and maintain a new ownership culture of creation processes.

Digital technology offers the concepts and the toolkit to prototype solutions and implement innovative ideas. The role of technology in the area of sustainability is to develop and provide efficient solutions with environmental impact. For instance, circular economy is an emerging trend based on the reusability (partially or fully) of the final industrial products. Instead of wasting resources and negatively affecting the environment, circular economy advocates the practice of recycling and the usage of biomaterials, thus becoming a promising way of creating profit and at the same time promoting social responsibility.

From an educational perspective, the successful implementation of the proposed pedagogy in higher education and the acquisition of the necessary 21st-century skills by the students in the area of sustainability require a number of carefully designed phases: (a) create interdisciplinary student-teams consisting of three to five members in order to foster collaboration; (b) allow students to search, find, and choose a real-world sustainability issue in the area of their own specialization (e.g., humanities, economics, social sciences, technology, and combinations); (c) collaborate with established enterprises, organizations, or NPOs (nonprofit organizations) to form a community of interest; (d) communicate with potential users with empathy and understand their preferences, beliefs, feelings, and culture; (e) co-design, co-create, and prototype a solution by using art-based practices and digital tools (Bjorklund et al., 2017); (f) document with photos, videos, and text each stage of the process and share the "WHY" students tackled with such an issue, "HOW" they managed to create a solution, and "WHAT" the solution is about, following Simon Sinek's "Start with Why" (Sinek, 2009); (g) exploit your design by taking an entrepreneurial risk, without the fear of failure (Long, 2012).

The method described above is a digital technology-enhanced, education-tailored approach that builds on design thinking (Brown, 2009) and double diamond methodology (Design Council, 2005).

Design thinking is the process of designing experiences with the active participation of the users and potential customers of the final product or service. It is mainly a human-centric approach that inspires innovation (Seitamaa-Hakkarainen et al., 2010; Leinonen and Durall, 2013). Linus Pauling once said that "The best way to have a good idea is to have lots of ideas." That is at the heart of design thinking; designers together with the users go through a process of iterations for exploring a range of potential solutions, getting feedback from the users, refining the idea, and thus moving towards the solution that better fits to user needs. Divergence and failure in the initial phase lead to convergence and success at the end of the process. During the initial process of understanding the real problem and users' needs, empathy is a skill of great concern, since user-oriented processes are usually ill-defined. On the other hand, brainstorming and collaboration between team members, mentors, and businesspeople as well as the implementation of a final prototype is also a challenging task. Digital competencies and art-based practices provide the means for realizing and applying innovative solutions. There are three fundamental criteria for classifying an idea as "potential solution": (a) be technically feasible, (b) be economically viable, and (c) be desirable to the users. However, prototyping an idea is no longer enough. Moving from the project room to the market is a great challenge that transforms researchers into young entrepreneurs. Being able to transform user needs into demand is a competency of extreme importance. Since the design is a human-centric process that is intended to address user needs in the form of a service or product, good design is heavily based on the vision of entrepreneurship (Kurani, 2017).

8.2 CONCEPTUAL FRAMEWORK AND LITERATURE REVIEW

John Donne once said, "No man is an island, entire of itself; every man is a piece of the continent, a part of the main; if a clod be washed away by the sea, Europe is the less, as well as if a promontory were, as well as if a manor of thy friend's or of thine own were; any man's death diminishes me, because I am involved in mankind. And therefore never send to know for whom the bell tolls; it tolls for thee." (Donne, 1624).

The Declaration of the Council of EU in Rome (Rome, 2017) stated that "In the ten years to come we want a Union that is safe and secure, prosperous, competitive, sustainable and socially responsible, and with the will and capacity of playing a key role in the world and of shaping globalization. We want a Union where citizens have new opportunities for cultural and social development and economic growth."

The goals set by EU in Rome are part of a wider agenda in the area of sustainable development, which involves a great number of difficult-to-be-solved problems such as energy overconsumption, water and food waste, nonoptimal use of electrical machinery, usage of non-ecological materials in everyday life (e.g., use of plastic), nonoptimal delivery methods used in logistics, extreme poverty, health issues, and many other fields of concern. The United Nations Educational, Scientific and Cultural Organization (UNESCO) has identified 17 areas of Sustainable Development Goals (SDGs) with 169 targets that need to be addressed over the next 15 years covering all the three dimensions of sustainability: economic, social, and environmental (UNESCO SDGs, 2017; United Nations 2030 Agenda for Sustainable Development, 2019).

The OECD Education Ministers (Meeting of the OECD Education Ministers, Paris, 3–4 April 2011) mentioned that "Sustainable development and social cohesion depend critically on the competencies of all of our population – with competencies understood to cover knowledge, skills, attitudes, and values" (OECD, 2011).

Skills-based educational frameworks are thought to be key enablers for the future of education (OECD, 2018a; Hughes and Acedo, 2017; UNESCO, 2013).

There are several frameworks for exploring the necessary skills and competencies for 21st century. The OECD DeSeCo (Definition and Selection of Competencies) framework considers three categories of competencies: use tools interactively, interact with heterogeneous groups, and act autonomously (OECD, 2005).

The OECD PISA Global Competencies framework (OECD-PISA, 2018b) considers global competencies as a way to investigate intercultural problems and allow people from different cultures to come closer and act collectively for achieving well-being and sustainable development.

The P21 framework (Partnership for 21st Century Learning (P21)) (P21, 2019) describes the competencies that the young adults must master in order to be success-ful in the workplace and their life. Apart from a basic set of key skills that are defined such as world languages, arts, and others, the updated 2019 version of the framework describes a number of many high-level interdisciplinary skills such as global aware-ness, entrepreneurial thought, environmental literacy, and innovative skills.

The World Economic Forum framework (World Economic Form and Boston Consulting Group, 2015) refers to the competencies that are needed in the future world market, the 21st-century marketplace. It considers several skills and compe-tencies, especially information and communication (ICT) skills, critical thinking and creativity, and also curiosity and decision-making.

The Council of Europe Competencies for Democratic Culture (Council of Europe, 2016) focuses on the skills that are needed for participating in democratic societies, especially intercultural competencies.

The UNESCO intercultural competencies framework (UNESCO-Competencies, 2013), also, refers to competencies needed for a globalized world. Especially, consider a mix of related competencies, such as communicative competencies and cultural competencies.

There are also frameworks for digital and entrepreneurial skills. Advanced digital skills are essential skills in 21st century and have been presented in several frame-works such as European Commission's – Digi Comp 2.0 and 2.1 (2016a and 2017), OECD Skills for a Digital World (OECD, 2016), UNESCO – Managing Tomorrow's Digital Skills (UNESCO-SKILLS, 2018), OECD-G20 – Key Issues for Digital Transformation (OECD-G20, 2017), LinkedIn – The Digital Talent Gap (Capgemini and LinkedIn, 2017), and European Parliament – Digital Skills in the 21st Century (European Parliament, 2018).

In the European Commission's DigiComp 2.0 and 2.1 (2016a and 2017), digital skills are placed in the framework of developing digital applications and content. In this framework, digital skills are considered key and advanced competencies for problem-solving and professional skills in conjunction with creativity, innovation, and entrepreneurship.

The OECD Skills for a Digital World Framework (OECD, 2016) considers that basic digital skills are no more enough and that graduates need programming skills in modern areas such as machine learning and others.

UNESCO's – Managing Tomorrow's Digital Skills (UNESCO Digital Skills, 2018) states that advanced digital competencies should be considered an essential part of an upskilling and reskilling process of professionals in order to be competitive in the labor market.

OECD and G20 (2017), in a joint report (January 2017), state that we need new approaches in the educational process that will aim in reskilling and upskilling of the workers in areas that are mostly used in the market and economy.

The Digital Talent Gap Report of Capgemini Research Institute and LinkedIn (Capgemini and LinkedIn, 2017) states that the digital gap is getting bigger and affects the strategic decisions of the enterprises regarding digital transformation.

Entrepreneurial skills, on the other hand, should also be considered essential 21st-century skills. EntreComp: The Entrepreneurship Competence Framework

(European Commission, 2016b) explores entrepreneurial skills and makes suggestions on educational practices and goals for entrepreneurial education. EntreComp was developed by the European Commission as a tool to improve the entrepreneurial capacity of European citizens and companies. Entrepreneurship is the process of creating cultural, social, or economic value. This framework defines the entrepreneurial competencies as a bridge between education and work and as a vehicle for individuals to actively contribute to economic development. Entrepreneurship as a competency involves decision-making; working with others; creating value; and including skills for driving transformation, innovation, and growth. Entrepreneurs should be able to understand the market opportunities and act boldly without limited by the fear of failure. A precondition for the successful introduction of entrepreneurial competencies in higher education is the active engagement of the youth in the process of continuous innovation (Allen and Clouth, 2012; Youssef et al., 2017; Franceschini et al., 2016).

However, innovative and entrepreneurial skills in the area of sustainable growth are not sufficiently offered by the current higher education system (Lozano et al., 2013).

In order to cultivate entrepreneurship for sustainability, the young higher education students and potential future entrepreneurs should be aware of the UNESCO's 2030 Agenda for Sustainable Development (UNESCO 2030 Agenda, 2015).

UNESCO considers 17 SDGs in their 2030 agenda for sustainable development with 169 targets that have to be addressed over the next 10–20 years.

Goal 1, ending poverty, is one of the most important goals. Poverty is usually measured as the number of people living with <$1.90/day. The number of extremely poor people has increased from 736 million in 2015 to 783 million in April 2018. About half of the total, 368 million people, live in five countries: Bangladesh, Democratic Republic of Congo, Ethiopia, India, and Nigeria (Atamanov et al., 2018; World Bank, 2019; Gapminder Tools, 2019).

Goal 2 is about ending hunger, achieving food security, improving nutrition, and promoting sustainable agriculture. Malnutrition, agricultural research and productivity, sustainable food production systems, proper function of agricultural markets, plant and livestock gene banks, investments in rural economy, ability to adapt to climate change, and food waste are some important areas of concern for achieving this specific goal (Tscharntke et al., 2012; Charles et al., 2014)

Goal 3 is about ensuring healthy living and promoting well-being for all at all ages by 2030, is another goal of extreme concern. Unfortunately, the mortality rates of children under the age of 5 range from 126/1,000 births in Somalia to 1.95/1,000 births in Finland [15, 16]. Life expectancy, which is the number of years we would expect a newborn child to live, ranges from 84.2 years in Japan to 51.1 in Lesotho. In Europe, life expectancy ranges from 83.5 years in Switzerland to 71.1 years in Russia (World Health Organization, 2018; UNICEF, 2018; World Bank Life Expectancy, 2019).

Goal 4 aims to ensure inclusive and equitable access to education and lifelong learning for boys and girls, adults, men, and women. By 2030, we should ensure that all learners will acquire all the necessary skills in order to promote sustainability (Webb et al., 2017; Milana et al. 2016; Vladimirova and Le Blanc, 2016).

Goal 5 is about achieving gender equality and empowering all women and girls. Discrimination against all women and girls should be ended, while violence and bullying should also be eliminated. Equal participation and access of girls and women in technology, political and public life, and also entrepreneurial education is also an important goal since there are still many inequalities in this area (EFA, 2012).

Goal 6 aims at sustainable management of water and sanitation for all. In 2015, 71% of the total world population (5.2 billion) used safe drinking water. Eighty-nine percentage of the total world population (6.5 billion) used a basic service (drinking water collector), whereas 844 million people lack the basic drinking water service (World Health Organization, Drinking Water, 2019).

Goal 7 is about affordable, reliable, sustainable access to energy for all. Until 2030, it should be guaranteed the universal and reliable access to energy services. The most important is to significantly increase the percentage of renewable energy resources in the global energy supply (McCollum, 2018).

Goal 8 targets at inclusive and sustainable economic growth, employability, and work for all. Especially, the goals for decent work for all, and for productive and high-quality employment, are extremely important goals for improving democracy, liberty, and social equity (International Labour Organization, 2018).

Goal 9 is about developing resilient infrastructures, promoting sustainable industrialization, and fostering innovation. These are the necessary factors for achieving sustainable development (DESA, 2014).

Goal 10, a goal that is related to the previous goals, is about the reduction of inequalities within and among countries. It involves income growth of the poor segments of the society, those with disabilities, and the less privileged sections of the society (Pan-African Network, 2009).

Goal 11 is related to everyday life standard of living, which is perhaps the most important factor for happiness. Making cities and human settlements inclusive, safe, and sustainable promotes current and future sustainable development (SDG Goal 11 Monitoring Framework, 2016).

Goal 12 tackles with two of the greatest problems in the area of sustainability: production and consumption. This goal states that consumption should take into account short-term, medium-term, and long-term restrictions, especially intergeneration considerations related to natural resources consumption. The reduction of food waste through recycling is a good practice and is important to encourage people and companies to adopt such practices (Fankhauser and Jotzo, 2018).

Goal 13 is about addressing the impact of climate change. Climate change is perhaps the most dangerous scenario for the planet's health that might lead to natural disasters. Educational policies for individuals and national policies for carbon reduction could promote mechanisms for addressing the impact of natural phenomena (Knight and Schor, 2014).

Goals 14 and 15 concern sustainability of the ocean and land life. Overfishing, destructive fishing, and marine pollution are some of the primary hazards. Sustainability of the land life is of equal importance as ocean life. Mountains, forests, biodiversity, and water ecosystems need protection based on sustainable policies that mainly aim to sustain livelihood opportunities and restore land ecosystems (Wackernagel et al., 2002).

Goal 16 is another fundamental goal for sustainable development. It is about creating peaceful societies with justice for all, including the marginalized segments of society, and strong institutions. According to this goal, reducing violence, exploitation, offensive behavior against human life and dignity, corruption, and organized crime promotes safety, democracy, and inclusive participation (The profile of SDG 16, 2019).

The final goal 17 is about promoting the implementation of the above objectives by promoting partnerships and enabling the necessary resource mobilization especially to developing countries. For instance, public debt sustainability is attained through restructuring; the standard of living convergence between countries with different growth rates through technology transfer, especially ICT and foreign direct investments; open trade system; political and macroeconomic stability; elimination of corruption; and moreover, reliable data. Technology plays a crucial role in achieving this goal. In particular, the use of ICT should be enhanced and thus promotes innovation and knowledge sharing (GUNi, 2018).

The above areas in sustainable development are going to impact the world's standard of living in the next decades. They need to be addressed in an efficient, collaborative, and innovative way since traditional practices have been proven to be inefficient or failed. The failure to tackle sustainable development is primarily a failure of the current educational and schooling system, which does not sufficiently offer the skills that are necessary for cultivating innovation, collaboration, personal and societal responsibility, and entrepreneurship (Michelsen and Wells, 2017).

8.3 METHODOLOGY

In this section, we analyze the methodology and the data received by 16 students organized in four teams to address real-world problems in the area of sustainability that was given as part of their coursework in the module "Digital educational content creation and usage, in contemporary learning methodologies," seventh semester, Teacher Education Program of Athens University of Economics and Business (TEP-AUEB) (www.dept.aueb.gr/en/tep), and "Educational Software Development," MSc in Digital Humanities, AUEB.

The subjects of the projects that were assigned to the students were in the areas of health, economics, social entrepreneurship, and cultural heritage. More precisely, the first team consisted of five undergraduate students; the project title was "Dietary Compliance for Diabetic Individuals." The second team consisted of four undergraduate students, and the project title was "Fruits Waste and Social Entrepreneurship." The third group consisted of four postgraduate students, and the project title was "Digital Museum Application." Finally, the fourth group consisted of three students, and the project title was "Vollunteera : The New Era of Volunteering."

Initially, the students formed smart teams and find a real-world issue in the sustainability area. After having found their team members, the students started to explore, investigate, and organize information found on the World Wide Web, in order to compile all the information and dimensions of their project. After that, the students should collaborate with supporting organizations, such as NPOs or other market agents, in order to extend their community and develop a wider community of common interests. Collaborating with enterprises gave the opportunity to the students to come closer to the real-world problems and entrepreneurial practices.

During the second phase of the project, the students developed a Gantt chart and the time schedule of their planned work. Moreover, the teams should create a mind map in order to visualize their thinking. Visualization of information is a nonlinear and richer way of description of the project. One great challenge for the students was to communicate with users and analyze user needs with interviews, observation, and empathy. The students were asked to make interviews with a diverse range of users such as average users, extreme users, and users that are indifferent about the services or products under consideration. At this stage, users' participation in the design process is a very important factor for co-creation of the product or service. Each team should get through three meeting cycles with the users. The iterations are important since possible solutions should be reduced based on users' feedback. At the same time, if the issue under investigation was not in line with users' perspective and preferences, then, the project brief should be rewritten and the issue should be redefined.

The team members were also asked to arrange a workshop with their community members such as supporting enterprise or organization in order to discuss the different perspectives and possible solutions.

The students were also asked to put emphasis and follow a journalistic, documentary process and document their research with media files such as photos, text, video, and audio in order to develop a multimedia version of their story and disseminate their experiences and ideas. Special attention was given to GDPR (General Data Protection Regulation) (GDRP, 2016).

The final phase involves the exploitation of the idea. Exploitation phase involves the development of entrepreneurial skills and spirit. Promoting the idea is heavily based on "why" the students chose this specific problem. This particular "why" should be at the heart of their exploitation and promotion phase.

8.4 RESULTS AND DISCUSSION

The data analysis builds on the work of Androutsos and Brinia (2019). The same methodology and questions are applied to a wider audience of 16 BSc and MSc students. The results confirm the findings of the previous study. The data analysis is based on the opinion of the students that run the methodology as a result of a Likert scale applied to a questionnaire with ten questions. The initial four questions are based on DARPA (Defense Advanced Research Projects Agency) Hard Test (DHT) (Luiz Fernando et al, 2018) in order to deduce the innovation level of students' projects and demos. The DHT is a test used at DARPA to evaluate the technology-based innovation of a project and has since been adopted as an innovation tool for team-based projects. The next six questions measure the impact of the pedagogy on students' acquired skills. We have set indicators that are used for measuring innovative, collaborative, and co-creative skills in comparison with legacy pedagogies. The students were asked to answer and rate the questions on a Likert scale from 1 to 7 (1 – I do not agree at all, 7 – I totally agree). The number of students who answered the questionnaire was 16 (females). The scale was based on DARPA scaling and the skills acquired level was considered "high" if the score was between 5 and 7 and "low" if it was between 1 and 4.

The questions posed to the students were the following:

Does the final solution require a paradigm shift in how it is considered across society?

Does the final solution require major advances in technical knowledge?

Does the final solution require multiple, distinct bodies of knowledge?

Does the final solution require little effort to begin taking action towards the solution?

Does the innovative index affect positively by the collaboration between team members?

Does the innovative index affect positively by the co-creation between team members and the users?

Have you acquired higher innovative skills in comparison with other teaching methods?

Have you acquired higher collaborative skills in comparison with other teaching methods?

Have you acquired higher co-creative skills in comparison with other teaching methods?

Have you acquired a greater motivation to be innovative in comparison with other teaching methods?

The results are depicted in Figure 8.1.

For each question, we have calculated the average score on the basis: "high" if the score was between 5 and 7 and "low" if it was between 1 and 4. We have also calculated the weighted average of the score in each question. The results are presented in Table 8.1.

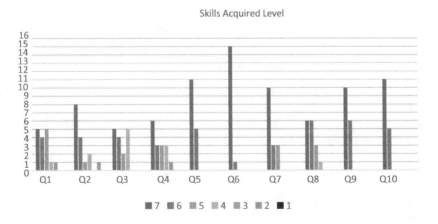

FIGURE 8.1   Number of answers per scale level (1–7) to each one of the ten questions.

TABLE 8.1
Results from the Joint DARPA/Likert Scale Test

Question Number		DARPA Test Score		Weighted Average
		Low (1–4) (%)	High (5–7) (%)	
DHT	Paradigm shift across society	12	88	5.6
	Major advances in technical knowledge	19	81	5.9
	Multiple, distinct bodies of knowledge	31	69	5.5
	Little effort needed to take action towards the solution	25	75	5.6
Collaboration and innovation		0	100	6.6
Co-creation and innovation		0	100	6.9
Innovative skills acquired vs other methods		0	100	6.4
Collaborative skills acquired in comparison with other methods		6	94	6
Co-creative skills acquired vs other methods		0	100	6.6
Motivation to be innovative in comparison with other methods		0	100	6.6

The objective of this study was to measure the innovation index of the project with respect to DRAPA (Q1–Q4), as well as the collaborative, co-creative, and innovative skills acquired by the students (Q4–Q10).

According to the proposed pedagogy, first and primarily the above indices determine the engagement of youth in innovative projects of sustainability relevance and the acquired skills in comparison with other pedagogies.

Moreover, in comparison with legacy pedagogies (Q7–Q10), this newly introduced pedagogy provides collaborative, co-creative, and innovative skills more efficiently than the legacy approaches, a fact that stresses the importance of the proposed pedagogy and encourages us to continue further on improving the proposed methodology.

8.5 CONCLUDING REMARKS

The proposed pedagogy introduces a method for teaching sustainability in higher education. Teaching sustainability in the 21st century requires innovative, collaborative, and co-creative teaching methods based on designedly ways of thinking.

The pedagogical model offered the opportunity to the students to see themselves in the framework of an extended community of common interests, to come close to the users in an interactive dialogue, to have the ownership of the learning process, to learn from peers through team collaboration, to understand the complex nature of entrepreneurship, to identify new opportunities for social value, to use digital tools for prototyping their idea and media tools for producing their digital story, and finally to prepare themselves to address with creativity complex real-world problems arising from a rapidly changing society and economy.

The proposed pedagogy encourages higher education students to take initiatives and participate in the design of their own future, becoming owners of their own lives. Being innovative is not a privilege of geniuses. It is the outcome of an experiential way of thinking and making. It is also the result of being creative. Creativity is closely related to art-based practices. TEP-AUEB promotes art-based practices by allowing TEP-AUEB students to be actively engaged in the production of art-based artifacts in collaboration with Athens School of Fine Arts. A sample of art-based activities can be found here: www.dept.aueb.gr/en/node/17412/.

REFERENCES

Allen, C. and Clouth, S. (2012). A Guidebook to the Green Economy, UN Division for Sustainable Development.

Androutsos, A., Brinia, V. (2019). Developing and Piloting a Pedagogy for Teaching Innovation, Collaboration, and Co-Creation in Secondary Education Based on Design Thinking, Digital Transformation, and Entrepreneurship. *Educ. Sci.* 9, 113.

Atamanov, A., et al. (2018). Global Poverty Monitoring Technical Note. Available Online: http://documents.worldbank.org/curated/en/173171524715215230/pdf/125776-REVISED-PUBLIC-CombineSpringWhatsNewfinal.pdf (accessed on 30/09/2019).

Bjorklund T. A., Laakso, M., Kirjavainen, S. and Ekman, K. (2017). Passion-Based Co-Creation. Aalto University.

Brown, T. (2009). Change by Design: How Design Thinking Transforms Organizations and Inspires Innovation. HarperBusines.

Capgemini and LinkedIn. (2017). The Digital Talent Gap.

Charles, H., Godfray J. and Garnett, T. (2014). Food security and sustainable intensification. *Philosophical transactions of the Royal Society B* 21(42), 107–116.

Council of Europe. (2016). Competencies for Democratic Culture: Living Together as Equals in Culturally Diverse Societies.

DESA. (2014). United Nations Department of Economic and Social Affairs (DESA). Population Division, World Urbanization Prospects: 2014 Revision, Highlights. ST/ESA/SER.A/352.

Design Council. (2005). The Design Process: What Is the Double Diamond? www.designcouncil.org.uk/news-opinion/design-process-what-double-diamond.

Donne, J. (1624). For Whom the Bell Tolls. Meditation XVII. Devotions Upon Emergent Occasions.

Durao, L. F. C. S., et al. (2018). Divergent prototyping effect on the final design solution: the role of "Dark Horse" prototype in innovation projects. *28th CIRP Design Conference*, Nantes, France.

EFA. (2012). EFA Goal 5: Gender Equality. UNESCO Office Bangkok and Regional Bureau for Education in Asia and the Pacific.

European Commission. (2016a). Digcomp 2.0 the Digital Competence Framework for Citizens.

European Commission. (2016b). EntreComp: The Entrepreneurship Competence Framework.

European Commission. (2017). Digcomp 2.1 the Digital Competence Framework for Citizens.

European Parliament. (2018). Digital Skill in the 21st Century.

Fankhauser, S. and Jotzo, F. (2018). Economic Growth and Development with Low-Carbon Energy. Wiley Interdisciplinary Reviews. Climate Change.

Franceschini, S., Lourenço G. D. F. and Jurowetzki, R. (2016). Unveiling scientific communities about sustainability and innovation. A bibliometric journey around sustainable terms. *Journal of Cleaner Production* 127, 72–83.

Gapminder Tools. (2019). Available Online: www.gapminder.org/ (accessed on 30/09/2019).

GDPR. (2016). General Data Protection Regulation. European Parliament, Council of the European Union.

GUNi. (2018). Approaches to SDG 17 Partnerships for the Sustainable Development Goals. Global University Network for Innovation.

Hughes, C. and Acedo, C. (2017). Guiding Principles for Learning in the Twenty-First Century. UNESCO.

International Labour Organization. (2018). *Decent Work and the Sustainable Development Goals: A Guidebook on SDG Labour Market Indicators, Department of Statistics (STATISTICS)*. Geneva: ILO.

Knight, K. and Schor, J. B. (2014). Economic growth and climate change: A cross-national analysis of territorial and consumption-based carbon emissions in high-income countries. *Sustainability* 6, 3722–3731.

Kurani, D. (2017). Good Design as Entrepreneurship. *HuffPost*.

Leinonen, T. and Durall, E. (2013). Design Thinking and Collaborative Learning.

Long. C. (2012). Teach Your Students to Fail Better with Design Thinking. International Society for Technology in Education (ISTE).

Lozano, R., Lukman, R., Lozano, F. J., Huisingh, D. and Lambrechts, W. (2013). Declarations for sustainability in higher education: Becoming better leaders, through addressing the university system. *Journal of Cleaner Production* 48, 10–19.

McCollum, D., et al. (2018). SDG 7 Ensure Access to Affordable, Reliable, Sustainable and Modern Energy for All.

Michelsen, G. and Wells, P. J. (2017). *A Decade of Progress on Education for Sustainable Development: Reflections from the UNESCO Chairs Programme*. UNESCO: Paris. ISBN 978-92-3-100227-4.

Milana, M., Rasmussen, P. and Holford, J. (2016). Societal sustainability: The contribution of adult education to sustainable societies. *International Review of Education* 62(5), 517–522.

OECD. (2005). The Definition and Selection of Key Competencies (DeSeCo).

OECD. (2011). Meeting of the OECD Education Ministers. Paris.

OECD. (2016). Skills for a Digital World. OECD.

OECD. (2018a). Future of Education and Skills, Education 2030.

OECD. (2018b). Global Competency for an Inclusive World.

OECD-G20 (2017). Key Issues for Digital Transformation in the G20.

P21. (2019). The Partnership for 21st Century Skills, Framework for 21st Century Learning Definitions.

Pan-African Network. (2009). Pan African e-network: A Model of South-South Cooperation, February to April 2009. Africa Quarterly.

SDG Goal 11 Monitoring Framework. (2016). A Guide to Assist National and Local Governments to Monitor and Report on SDG Goal 11 Indicators. UNESCO, World Health Organization.

Seitamaa-Hakkarainen, P. Marjut Viilo, Z. and Hakkarainen, K. (2010). Learning by collaborative designing: Technology-enhanced knowledge practices. *International Journal of Technology and Design Education* 20, 109–136.

Sinek, S. (2009). *Start with Why: How Great Leaders Inspire Everyone to Take Action*. Portfolio: New York.

The Profile of SDG 16. (2019). Report. United Nations Development Programme (UNDP).

The Rome Declaration. (2017). Declaration of the Leaders of 27 Member States and of the European Council, the European Parliament and the European Commission. Council of the EU.

Tscharntke, T., et al. (2012). Global food security, biodiversity conservation and the future of agricultural intensification. *Biological Conservation* 151(1), 53–59.

UNESCO. (2013). Towards Universal Learning.

UNESCO 2030 Agenda. (2015). Transforming Our World: The 2030 Agenda for Sustainable Development. UNESCO.

UNESCO and Sustainable Development Goals. (2019). United Nations Education Science and Culture Organization.

UNESCO Digital Skills. (2018). Managing Tomorrow's Digital Skills: What Conclusions Can We Draw from International Comparative Indicators?

UNESCO-Competencies. (2013). Intercultural Competencies Framework.

UNICEF. (2018). Under-5 Mortality. Available Online: https://data.unicef.org/topic/child-survival/under-five-mortality/ (accessed on 30/09/2019).

United Nations 2030 Agenda for Sustainable Development. (2019). United Nations Education Science and Culture Organization.

Vladimirova, K. and Le Blanc, D. (2016). Exploring links between education and sustainable development goals through the lens of UN flagship reports. *Sustainable Development* 24, 254–271.

Wackernagel, M., et al. (2002). Tracking the ecological overshoot of the human economy. *Proceedings of the National Academy of Sciences of the United States of America* 99, 9266–9271.

Webb, S. et. al. (2017). Lifelong learning for quality education: Exploring the neglected aspect of sustainable development goal 4. *International Journal of Lifelong Education*, 509–511. Available Online: www.tandfonline.com/doi/full/10.1080/02601370.2017.139 8489 (accessed on 30/09/2019).

World Bank. (2019). Understanding Poverty. Available Online: www.worldbank.org/en/topic/poverty (accessed on 30/09/2019).

World Bank Life Expectancy. (2019). Life Expectancy at Birth. Available Online: https://data.worldbank.org/indicator/sp.dyn.le00.in (accessed on 30/09/2019).

World Economic Form and Boston Consulting Group. (2015). New Vision for Education: Unlocking the potential of Technology.

World Health Organization. (2018). Available Online: www.who.int/news-room/fact-sheets/detail/children-reducing-mortality (accessed on 30/09/2019).

World Health Organization, Drinking Water. (2019). Key Facts. Available Online: www.who.int/news-room/fact-sheets/detail/drinking-water (accessed on 30/09/2019).

Youssef, A. B., Boubaker, S. and Omri, A. (2017). *Entrepreneurship and Sustainability Goals: The Need for Innovative and Institutional Solutions.* Technological Forecasting and Social Change, Elsevier: Amsterdam.

Index